TALES FROM THE
WINE FLOOR

TALES FROM THE
WINE FLOOR

100 QUESTIONS
ASKED OF A SOMMELIER

James Quaile
Drawings by John O'Brien

LP
LYONS
PRESS

Essex, Connecticut

An imprint of The Rowman & Littlefield Publishing Group, Inc.
4501 Forbes Blvd., Ste. 200
Lanham, MD 20706
www.rowman.com

Distributed by NATIONAL BOOK NETWORK

British Library Cataloguing in Publication Information available

Library of Congress Cataloging-in-Publication Data

Names: Quaile, James, author. | O'Brien, John, 1953- illustrator.
Title: Tales from the wine floor : 100 questions asked of a sommelier / James Quaile ; illustrations by John O'Brien.
Description: Lanham, MD : Lyons Press, an imprint of The Rowman & Littlefield Publishing Group, Inc., [2023] | Includes bibliographical references and index.
Identifiers: LCCN 2023007673 (print) | LCCN 2023007674 (ebook) | ISBN 9781493074655 (hardcover) | ISBN 9781493078486 (ebook) Subjects: LCSH: Wine and wine making—Miscellanea.
Classification: LCC TP548 .Q35 2023 (print) | LCC TP548 (ebook) | DDC 663/.2–dc23/eng/20230302
LC record available at https://lccn.loc.gov/2023007673
LC ebook record available at https://lccn.loc.gov/2023007674

For my girls
Janet, Lynsay, Leah, Aven, Shea, Ailidh, and Annabel

Contents

Chapter Three
BUYING

Chapter Four
PAIRING & SERVING

Chapter Five
SAVING

Chapter Six
STORING

Chapter Seven
NAMING

Chapter Eight
FORTIFYING AND SPARKLING

Chapter Nine
READING

Chapter Ten
TRAVELING

Introduction

Questions, questions, questions. If you are a wine professional, you are used to answering questions. The winemaker is asked about weather and harvest. A wine distributor is asked about availability and price. As for me, a Certified Sommelier/wine consultant, I spend my day answering questions about, well, everything. Wine can be fun, and sometimes people have funny questions about it.

I began compiling my favorite customer questions and posting the most amusing ones online. I called them my "Tales from the Wine Floor." As these tales grew in number, I began to realize that the answers I thought were clearly obvious weren't so obvious at all. Some were even insightful. Why is wine stored in oak barrels? Who started that? Is bad wine harmful? Hmm. Why don't wine labels list ingredients? Good question!

I am Jimmy Quaile. I wasn't born in the South of France and raised on a vineyard by a family of sommeliers. I was raised in a working-class family with 10 siblings in a tiny house in New Jersey. I didn't even drink wine until I was well into my 30s. Having spent most of my life onstage as a drummer and singer, the prospect of becoming a Certified Sommelier (and writing a book about wine) was not even on my bingo card.

My earliest wine memory is Thanksgiving dinner when my mother would allow me (and my 10 brothers and sisters) to have a small sip of Harvey's Bristol Cream, her favorite. The clinking of glasses for a toast seems so adult and sophisticated when you're a kid. To this day, I can't walk past those blue bottles on the shelf without thinking of my mom. That's what wine is all about—the people, the places, the memories.

After those early memories, my next wine encounter was less about feeling sophisticated and more about feeling, well, anxious. My wine anxiety arose, as I believe it does for many, with the presentation of an intimidating wine list at a hoity-toity restaurant.

"Sir," said the tuxedo-clad waiter, "the wine list."

I can play this game, I thought. I sat up straight, held up my pinky, and opened the leather folder. That's when my posture began to slouch. *Wait. Where are the descriptions?* I wondered. The food menu overflowed with descriptors: center-cut black Angus beef tenderloin baked in a puff pastry with wild mushroom duxelles; lump crabmeat over pistachio-encrusted Norwegian salmon. Every dish was explained in minute detail. But the wine list? Nothing but names that I couldn't pronounce and prices that I couldn't afford. How was the average Joe—or the average Jimmy, in this case—supposed to know what to order? Where was the wine cheat sheet?

Then I got to thinking. Why *does* wine have to be so damn complicated? The unpronounceable names, the exorbitant prices . . . it's grape juice for Pete's sake! And then there were the "somms." In the schools I attended, I'd never even heard of a sommelier. So it just seemed weird to be taking recommendations from a guy with a flat silver cup around his neck.

After that anxiety-inducing memory, my attitude toward wine took a positive turn. It occurred when I visited Paris for the first time with my wife and two daughters. Walking along the Champs-Élysées, we stopped and shared a carafe of Beaujolais in a quaint bistro. The fruity and uncomplicated wine was fantastic. When I got back home, I picked up the exact bottle. It was good but didn't quite taste the same. It turns out wine tastes a lot different when you drink it in a bistro on the Champs-Élysées than when you drink it in your kitchen in New Jersey.

That lesson sparked my curiosity about wine. I started trying all different kinds of wine. The more I tasted, the more I loved it. I was hooked. I began buying wine books and highlighting the important facts. I researched regions and grapes, and I tasted as many bottles of wine as I could afford. After several years of this self-guided wine school, I landed my dream job as "the wine guy" at Roger Wilco, a large retail wine store in New Jersey. I spent every day choosing wines, tasting wines, and answering questions about wine. Eventually I wondered if I knew enough to become a Certified Sommelier. That, by the way, is no small task. The test is given by the Court of Master Sommeliers a few times a year and costs over $1,000. If you fail, you have to wait to retake it, travel to whichever city is giving the test at the time, and repay for it. And it's considered to be one humdinger of a test.

"You could write a book on this stuff," said my friends.

"You should do it!" said my coworkers.

"You better not fail," said my wife.

I analyzed the situation and came to a conclusion: I couldn't lose. I don't mean it was a win-win situation. I mean, literally, I couldn't lose that amount of money with nothing to show for it, and I couldn't lose that much time without a paycheck. And so, being the practical person that I am—and thinking that flat cups worn around the neck (called *tastevins* don'tcha know) are actually an impressive fashion statement—I decided to take the ridiculously pompous test. I did study my ass off, but, really, no one was more shocked than me that I passed! (I explain what that crazy experience was like at the end of the book.) After passing the test, I was over the moon. *This is it!* I thought. *I'm a Certified Sommelier! I'm in the big leagues! I'm going to attend the most exclusive events! People will bow to my wine expertise!*

And then I went back to my job on the wine floor, where reality slapped me in the face.

"Can you tell me where I can find the 'Chippendales'?" a customer asked.

"You mean the Zinfandels?" I answered.

"Oh, yeah, that."

Next customer: "Can you tell me where to find the wines that don't give me a headache?"

I felt like screaming, "Don't you see my fancy lapel pin? Do you know how hard I studied to get this?" I didn't do that, however, because I realized that the vast majority of customers don't really give a damn about wine vintages and appellations—or a CMS pin for that matter! Mostly they just want me to tell them what to bring to Aunt Flossie's birthday party where no one really cares which wine pairs with what food. That's when it occurred to me. *I* should be the one who writes a wine book for the average Joe (and Jimmy)!

And so here is my book: a compilation of the basic—sometimes absurd, sometimes hysterically funny—questions asked by regular wine-drinking people. And the answers didn't come from someone raised in the South of France by a family of sommeliers. It's a book about wine, told in a way that I wish someone had told it to me.

I hope reading this makes you want to try anything and everything. And don't think for a second that you don't have a "sophisticated palate." That's nonsense. Wine is for everyone to enjoy, not just self-proclaimed wine connoisseurs or snooty sommeliers. Enjoy your life. Wine will go a long way in making that possible.

My wish is that you'll read this book and say, "*Bingo!*"

Tale from the Wine Floor

Customer: 'Scuse me, can you tell where your White Zinfandels are?

Me: Sure, come this way. Here are all of our White Zinfandels.

C: Don't you have the white one? I only like white wines.

M: Well, I can show you other white wines, but these are White Zinfandels.

C: OK, then I'll have a Pinot Noir.

M: I'll show you where they are, but Pinot Noir is a red wine.

C: No, I know they make a white Pinot Noir.

M: I'm afraid you're thinking of Pinot Grigio. That's a white wine.

C: Is there somebody else I can talk to that knows wine?

1. WHAT IS A SOMMELIER?

Sommelier (pronounced SUH-muhl-yay) may seem fancy and difficult to pronounce, but it actually just means "butler." It is a derivative of the French word *sommier*—one who arranges transport. In England during the 1300s, the sommelier was a royal butler who would source wines for the king. In modern times a sommelier is a person who specializes in all aspects of beverage service, including wine-and-food pairing, beer, sake, and classic cocktails. It used to include cigars and water(!), but those categories have been dropped. In the wine industry, there are several organizations that offer certifications. The Court of Master Sommeliers, of which I am a part, was established in the United Kingdom in 1977. It has four levels: Intro, Certified Sommelier, Advanced Sommelier, and Master Sommelier. The exam for the certification level has three parts: Blind Tasting (identify the grape, region, and vintage), Theory (a written exam), and Service (literally serving a table of guests). Only after passing this exam can hopeful wine connoisseurs officially wear the pin of a "sommelier." You could *call* yourself a sommelier, as some do, without being certified. But that's like saying you're a lawyer because you went to law school. You aren't a lawyer until you pass the bar.

To get a realistic peek at what it's like to take "the test," flip back to question #100 (especially if you're one of those "read-the-ending-first" types).

Tale from the Wine Floor

Customer: What is the thing they call you?

Me: 'Scuse me?

C: You know, a sominair.

M: Oh, a sommelier. Yeah.

C: I could be one of them. I love wine.

M: OK. What are some of your favorites?

C: I like all wine. Pink, white, red, I just love wine.

M: Good for you. You must have a broad palate.

C: Yep, as long as it says Moscato, I love 'em all.

Chapter One

MAKING

2. HOW IS WINE MADE?

"Wine is proof that God loves us and wants us to be happy." That phrase, popularized by bumper stickers, T-shirts, and memes, is attributed to Benjamin Franklin. While it is not true that Ben actually said that, the implication that wine was given to us by a higher power probably is. Humans didn't invent wine. Yeast occurs naturally on the grape skins as the grape grows. When the yeast comes in contact with the juice (by crushing), it eats the grape's natural sugar, and ta-da! You've got wine.

Humans did figure out how to move the winemaking process along and improve the final product in roughly 5000 BC (give or take a thousand years), but fermentation itself was and still is a mystery. It wasn't until 1857 that Louis Pasteur defined the biological process of fermentation. So, every winemaker before then had no idea what the hell was happening! They must've figured, "Hey, it works, and it tastes good—who cares why?"

The recipe for wine is pretty basic: pick the grapes, crush them, and put them into a container to ferment. But, as in any great recipe, the secret is in what winemakers use and how they use it. So, if you were thinking of buying a bunch of grapes from the supermarket, crushing them with your bare feet, and making your own wine, think again. For one thing, grapes used in winemaking are different from table grapes. Wine grapes come from the species *Vitis vinifera*, which is native to southern Europe. It accounts for 90% of the cultivated grapes in the world. There are literally thousands of different *Vitis vinifera* grape varietals (wine speak for varieties of grapes). Interestingly, all but a dozen or so grape varietals have white (or clear) inner flesh. The only difference between red, white, Rosé, or even Orange wine is how long the juice stays in contact with the outer skins. So, for those of you who think you only like red wine, in reality you only like white wine!

Back to the recipe. Aside from knowing which grapes to plant and where to plant them, the most important job for a winemaker is choosing the perfect time to harvest the grapes. The goal is for the sugar, tannin, acidity, and phenolic (aromatic) compounds to all reach maturity at about the same time . . . or as close to the same time as possible. There are tools to measure these levels chemically, but a great winemaker's instinct and experience will always prevail over the technical data.

After choosing the all-important harvest date, winemakers must then decide if they want to use the grape's natural ambient yeast or inoculate the juice with cultured yeast strains. Then nature takes over, and the yeast does what yeast does—consumes the sugar and releases ethanol, carbon dioxide (CO_2), and heat. The CO_2 and heat escape and the ethanol remains. The more sugars there are in the grapes, the higher the alcohol level of the wine.

Finally, winemakers choose a vessel to use for aging (stainless steel, concrete, ceramic tank, or wooden barrel). Every little step in the recipe makes the difference between an everyday wine that is consistent from vintage to vintage, and a one-of-a-kind work of art.

So, whether you believe that wine was made by the hand of God or the hands of man, clearly somebody wants us to be happy. Thank God.

Tale from the Wine Floor

Customer: Do you have that wine made with marijuana?

Me: The what?

C: It's a wine with marijuana in it.

M: What is it called?

C: I don't remember.

M: Where did you have it?

C: I forget.

3. HOW MANY GRAPES GO INTO A BOTTLE OF WINE?

It takes about 2.5 pounds of grapes to make a bottle of wine. The number of grapes isn't specific because there are a lot of variables to consider. Grapes with thick skins have less juice than thin-skinned varietals. And some grapes, like those grown in Champagne, are surprisingly small. Evaporation is also a factor. Winemakers use the French term "ullage" for that evaporation, which can be as much as 10%. Makers of Scotch and Bourbon refer to that volume loss as "angel's share."

Here is wine by the numbers:

75 grapes = 1 cluster
1 cluster = 1 glass
4 clusters = 1 bottle
40 clusters = 1 vine
1 vine = 10 bottles
30 vines = 1 barrel
1 barrel = 60 gallons

4. WHAT ARE TANNINS?

Tannins, technically called polyphenols, are naturally occurring molecules. They are found in plants, seeds, bark, wood, leaves, and fruit skins. In winemaking, tannins come from the skins, seeds, and stems of grapes, mainly red grapes, and they can also come from the oak barrels. Other foods also have tannins. Tea, walnuts, cinnamon, clove, and dark chocolate all have high levels of them. Tannins add more of a textural sensation than something that affects flavor, though some say that it makes the wine "taste" more astringent, or in wine lingo, dry.

The compound derives its name from the process of curing leather, or tanning! Ecologically, a plant (or grape) has tannins to make them unpalatable to animals that might otherwise eat them. In that way, they are the plant equivalent of a skunk's scent or porcupine's quills. Odd that they are prized in winemaking, but there are reasons to love tannins. Winemakers love them because they are an antioxidant that both protects the wine and adds structure to their final product. Wine collectors love tannins because they make the wine age-worthy since tannins become softer and silky smooth over time.

On the other hand, sweet-wine lovers hate tannins. They even wonder why anyone in their right mind would like such a tongue-drying, bitter-tasting, mouth-puckering, chalky feeling in their mouth. To be sure, there is no middle ground on tannins. You either love 'em or hate 'em. If you're in the hate 'em camp, I have good news and bad news for you. The bad news is that most of the greatest wines in the world start with ample amounts of tannins. The good news is that tannins can be lessened or even neutralized by exposure to air. That's one of the main reasons for decanting red wine. For more on decanting, jump to question #50.

Tale from the Wine Floor

Customer: Can you help me pick a wine?

Me: I'd be happy to. Tell me about the style of wine you like.

C: Well, I don't like wines that make my face do 'dis.

M: I'm guessing you mean you don't like tannic or dry wines. Here is a very popular sweet wine I think you'll like.

C: Do you like it?

M: No. It makes my face do 'dis!

5. WHAT ARE SULFITES?

Because there is a warning label on wine bottles that says "contains sulfites," people assume that sulfites are bad. The truth is that they *can* be bad for a *very small* percentage of people, but for most of us, they are nothing to worry about. Let me explain.

Sulfur dioxide (also known as sulfite) is a chemical compound made up of sulfur and oxygen. It occurs naturally but can also be produced in a laboratory. Measured in parts per million (ppm), it has been used for thousands of years as an antioxidant and antimicrobial to preserve foods and beverages. As far back as the eighth century BC, ancient Greeks, who used sulfur to fumigate ships, noticed that food lasted longer in the treated room. In wine, it can stop fermentation at a specific time, act as a preservative, and prevent bacteria and oxidation.

Since sulfur dioxide is a natural by-product of yeast, the only way to produce a genuinely sulfite-free wine is to put it through chemical manipulations that strip the naturally occurring sulfur out of the wine. Wines from Italy, France, Argentina, Spain, Chile—or any wine-producing country for that matter—*all* contain sulfites. The only difference between these countries and the United States is that they do not require a warning label for their wines.

So why does the United States require the warning on the bottle? It is because of the salad bar! In the 1970s and 1980s, the salad bar gained popularity, and the fruits and vegetables were routinely sprayed with a high amount of sulfites—sometimes up to 2,000 ppm—to prevent the produce from wilting and turning brown. (Most wines contain less than 100 ppm.) The FDA received complaints from people having adverse reactions. Strom Thurmond, a teetotaling, anti-alcohol senator from South Carolina, lobbied to make the "contains sulfites" warning label part of the Anti-Drug Abuse Act of 1988, a continuation of the "War on Drugs."

Consumers have been confused ever since. To be clear, there are certainly people, 0.4% of the population, who are indeed allergic to the compound. However, this is not always to blame for some people's adverse reaction to it. For example, why do some people get a headache when they drink red wine? The latest research suggests histamines, the stuff of allergies, as the main reason. Food and drinks that have been aged, such as aged meats and red wines, cause our body to release histamines that can create allergy-type symptoms for some people. Taking an antihistamine an hour before drinking may reduce or solve the problem.

If you still believe that sulfites in wine give you a headache, try a little experiment: Eat some dried apricots. If you do not get a headache, sulfites are not the culprit. A two-ounce portion of dried apricots has *10 times* more sulfites than one glass of wine. Still not convinced? Try cookies, crackers, pizza crust, flour tortillas, pickles, salad dressings, olives, vinegar, shrimp, scallops, sugar (brown and white), fruit juice, deli meat, and even prescription pills. The point is, there is no medical research data that shows sulfites as the cause of headaches. If you're really looking to lessen your sulfite intake, though, know that white wines contain more sulfites than red wines, and sweet wines contain more sulfites than dry wines.

Note: Organic wine does not mean sulfite-free wine. It means there were no *added* sulfites.

For argument's sake, let's say you are one of the 0.4% who has a sulfite allergy but wants to drink wine. Is there a DIY hack that can remove sulfites from wine after they are in the bottle? It turns out there is! And it's probably in your bathroom cabinet right now: hydrogen peroxide. The theory is that a few drops of H_2O_2 in your wine will turn sulfite into hydrogen sulfate, thereby eliminating them altogether. Have I tried it? No. *And I don't recommend it.* But I don't seem to get headaches from wine—unless I drink too much. In which case, the fix is also in my bathroom cabinet. It's called Tylenol.

Tale from the Wine Floor

Customer: Where are the wines that don't give me a headache?

Me: You're in luck. We put all the wines that give people headaches in one section. Sorry, I couldn't help myself.

C: What?

M: I was kidding.

C: Oh, so you don't have a section for those?

6. WHAT EXACTLY IS SEDIMENT?

Sediment is the tiny sand-like particles that fall to the bottom of a bottle, tank, or barrel. It comes from the winemaking process or the aging process—or both. Dead yeast cells, fragments of the grape skins and seeds, tartrates, polymers, and any other fragments that are too heavy to stay soluble in liquid are the initial culprit. Winemakers are able to remove these particles before bottling using fining techniques that filter them out, but some believe that they give flavor and texture to the wine.

Sediment can also be the by-product of age. Molecules that were initially too small to be filtered can bind together over time—making them too large to stay suspended in the juice, so they fall to the bottom as well. That's why red wine gets lighter in color as it ages. Most white wine is filtered and clarified to produce a clear and shiny appearance. It has very little grape skin contact and therefore has less chance of "throwing off" sediment. White wine, however, can leave tartrate crystals, which are a different kind of deposit. Both sediment and tartrates, though gritty and unpleasant, are entirely harmless. They are so harmless that the leavening agent cream of tartar—which is probably on your kitchen spice rack—is made from wine tartrates. To me, it's much ado about nothing. After all, I see the same gunk at the bottom of my coffee cup and don't give it a second thought.

7. DO WINEMAKERS ADD FLAVORS TO WINE?

It may surprise some people that the answer is no. Winemakers do not add flavors to the wine. The flavors and aromas in wine occur naturally during the winemaking process. Winemakers can't put the juice from fruit other than grapes in the bottle unless they state it on the label ("plum-flavored wine," etc.).

Tale from the Wine Floor

Customer: I'm looking for the Pinot Noir that has the cherries in it.

Me: Well, Pinot Noir doesn't really have cherries in it, but some people think they can taste notes of cherry in the wine.

C: Well, they shouldn't say it tastes like cherries if there aren't any cherries in it. That's dumb.

M: Uh-huh.

C: So this chocolate wine doesn't really have chocolate in it?

M: Actually, that does have chocolate in it.

C: So how do I know which wines have what it says in it and which wines don't? That's dumb.

M: Funny, but that actually sounds like a good question!

C: Do you like this chocolate wine?

M: No. That's dumb.

8. HOW DO OTHER FRUIT FLAVORS GET INTO WINE?

The other "flavors" have another name. They are called esters. Esters are chemical compounds that are created during fermentation and have an aromatic, fruity quality. Esters are also found in other fruits and vegetables and are responsible for their aroma. However, rather than saying a Sauvignon Blanc has a methoxy-pyrazine character, we describe bell pepper and asparagus notes. Pinot Noir can remind you of cherry. Riesling can taste like somebody added green apple. And I could swear banana flavoring was added to Beaujolais Nouveau.

Still, other aromas and tastes have nothing to do with the grape. The barrel used and time in the bottle bring their own nuances to the mix by adding notes of leather, coffee, or cigar box. Now, do winemakers try to coax those certain flavors out of the grapes and into their wine? Absolutely. That is the art of winemaking.

You may wonder why the taste or smell of something like leather is appealing to wine lovers in the first place. That's when wine gets personal. It can trigger a memory, like oiling up your baseball glove at the start of summer or falling into your favorite lounge chair. Wine is more than a beverage—it's comfort in a glass. Granted, some descriptors make you shake your head, but "wine speak" is sometimes the only way to describe such a complex product.

9. ARE THERE ADDITIVES IN WINE?

Yes, there are. But before we dive into the world of additives, it should be stated that every product that is put into a container, stored for any length of time, and shipped someplace else has had chemicals added to it, and with good reason. Long, hyphenated chemical names ending with the letters "hyde," "thyl," or "ate" seem to trigger alarm bells for many who don't possess the scientific education to interpret such names. I am one of those without food science knowledge, but I can say with certainty that there are multitudes of hidden ingredients in commercial products, a major one being *an agenda*. Be sure to "consider the source" of claims of toxins/poisons and the like in wine. I've found that fear-mongering about additives usually is made to convince a buyer to buy something else, say, someone else's product—which also happens to contain additives!

Still, a discussion is warranted, so here goes.

Like alcohol itself, any chemical in its pure form is toxic and dangerous. But in parts per million (ppm) or even billion (ppb), they are deemed to be a safe alternative. As it pertains to wine, there are two main problems that chemical additives attempt to solve:

1. Wine needs to be free from bacteria for a reasonable amount of time, so a preservative/stabilizer is needed for fermenting, aging, and storing. That's a good thing.
2. Since most wine drinkers don't want particles floating around in their wine, a fining/clarification agent is needed. Particles that are big enough to see are easy to filter out (e.g., dead yeast, skins, seeds, stems). But if the particles are too small (and make the liquid hazy), there are chemicals that make them bind together so that they also can be filtered out. I am OK with clarification agents, but I can certainly see the arguments for leaving wine in its natural state.

And that's it. I am not advocating for the use of chemicals in anything. Of course, I would rather drink, eat, and use products that are naturally safe and require no additives. But I am a realist and understand that it is difficult in today's world to be chemically off the grid.

10. WHAT IS ORGANIC WINE?

"Organic" wine is made with organically grown grapes, meaning all additives are organic, and no genetically modified organisms (GMOs) or other prohibited ingredients are allowed. It also cannot have any *added* sulfites. Naturally occurring sulfites are permitted but cannot exceed 20 parts per million. Organic does *not* mean a wine is vegan. If you see a label that reads "Made with Organically Grown Grapes," sulfites may be added, though they cannot exceed 100 parts per million.

Note: The US government regulates the term "organic," but "sustainable," "biodynamic," and "natural" have no legal definitions.

Tale from the Wine Floor

Customer: I'm looking for a wine from anywhere except the United States.

Me: We have a lot of great choices . . . France, Italy, Spain, etc. . . . But I'm curious—why don't you drink wine from the United States?

C: Because we add sulfites. They give me a headache.

M: Actually, every country adds sulfites, but only the United States and Australia require "Contains sulfites" to be on the label. On top of that, if you were allergic to sulfites, you would get a lot of things, but a headache isn't one of them.

C: So why is it added in the first place?

M: It's a preservative.

C: Then why do I get headaches when I drink a wine from America?

M: Maybe it's the person you're drinking it with.

11. WHAT IS BIODYNAMIC WINE?

"Biodynamic" farming is based on the agricultural ideas and practices of the Austrian philosopher Rudolph Steiner. The basic concept of this kind of farming is that everything in the universe is interconnected, including the vine, humans, earth, and stars. It's an alternative method of agriculture. For example, in biodynamic farming, one of the more unusual practices is called Preparation 500. It involves taking a horn from a lactating cow, filling it with manure, and burying it 16 to 18 inches into the soil during the dormant months of winter. When the horn is dug up, the waste is mixed with water and sprayed over the vineyard four times a year according to the phases of the moon—and always in the afternoon. No, I'm not kidding. Does it work? Many well-respected winemakers think so. And you thought viticulture was boring.

12. WHAT IS SUSTAINABLE WINE?

According to the California Sustainable Winegrowing Alliance, a sustainable winery "incorporates both organic and biodynamic principles and practices, in addition to earth-friendly methods for vineyards and surrounding ecosystems, including energy efficiency, the protection of air and water quality and enhanced relations with employees and neighbors." Whew! In other words, it means they create a wine with little or no chemical intervention, leave the earth the way they found it, and are nice to people. Be kind and drink wine! Now that's a motto I can get behind!

13. WHAT IS NATURAL WINE?

The term "raw" wine would seem to be a more apt description, but "natural" wine is made using organically grown grapes that are harvested by hand and vinified using indigenous yeast. Although a minimal amount of sulfur dioxide is sometimes used, it is pure grape juice with nothing added and nothing taken away. That means no chemicals in the vineyard, no mechanical separation of the stems from the grapes, no punching down or pumping over. No nothing. Natural winemakers are the hippies of wine! Natural wines are also unfiltered, so you can expect them to have some sediment in the bottle and be a bit hazy in the glass. I know it doesn't sound very appealing, but devotees (and there are many) believe that you are tasting a living wine with its own inherent energy. I will say they are kinda funky and kinda fun. Kinda.

This hands-off approach to winemaking results in a much shorter shelf life. So, with all natural wines, the younger, the better.

Note: In March 2020 France officially recognized a new term for natural wines called *vin méthode nature*. No other country has any such denomination. Yet.

14. WHAT IS A PÉT-NAT?

Petulant by definition is either a word to describe your bratty little cousin or a French word meaning effervescent. Both apply when talking about *Pétillant Naturel* wines—or as millennial hipsters call them, "Pét-Nats." They are a wilder and more unpredictable version of sparkling wines. Although they have gotten a pop in popularity with the recent surge of natural wines in restaurants and wine bars, they are anything but new. The term used when describing how they are made is *méthode ancestrale*, the ancestral method of getting bubbles into the bottle. They simply bottle the juice before the fermentation is finished! This method predates the more modern *méthode champenoise* by over a hundred years. Its roots were traced back to 1531 in the Loire Valley in France.

Pét-Nat winemakers can use any grape (or grapes), so you can find red, white, Rosé, and even Orange Pét-Nats. They are typically low in sugar and alcohol. You can tell a Pét-Nat from other sparkling wines by the closure. Rather than a traditional cork and cage like Champagne, they will most often have a crown cap like a beer.

As for the flavor, proponents of the frothy and fruity wine love their rustic nature. Opponents think they are just weird.

15. WHAT IS ORANGE WINE?

As if wine weren't confusing enough, Orange wine is a white wine made the way red wine is made, using grapes that are actually green. The skins of the grape are left to ferment along with the juice. The wine's orange hue comes from the amount of time the juice spends in contact with the skins—this also adds tannins the same way it does for red wine. Orange wines tend to have a rich and spicy tang that can be intriguing but can also confuse even the most knowledgeable wine lover. This winemaking practice dates back hundreds of years in Slovenia, thousands of years in the eastern European country now called Georgia (under the name "amber wine"), and was even fairly common in Italy in the 1950s and 1960s. The term "Orange Wine" is attributed to British wine importer David A. Harvey, in 2004. Before that it was called, uh, wine.

16. WHAT IS KOSHER WINE?

The Hebrew word "kosher" literally means "fit." The foods that are fit for consumption for a Jew are a part of kosher law. According to Judaism, these laws were commanded by God and handed down to the children of Israel in the Sinai Desert by Abraham. To a Jew, holiness is not confined to holy places—they believe that all of life is a sacred endeavor. So, eating and drinking are godly acts.

Fruits, vegetables, and grains are kosher in their natural state. But since wine was served during the services in the Holy Temple, Torah law requires it to be produced and handled by Torah-observant Jews who must follow kosher rules to a T . . . or in this case, a P. The initial P on a kosher product stands for Passover. You will also see an OU on a package—designated by a U inside an O—meaning the Orthodox Union. This certification guarantees that the product was overseen by a Sabbath-observant Jew or rabbi who ensured that it was made in accordance with Jewish Dietary Laws (Kashrut). For those who want even more insurance, a wine can be "mevushal," meaning cooked or boiled. A mevushal wine keeps the wine kosher even if touched by a gentile (non-Jewish). Here is the reasoning: Thousands of years ago, wine was served by pagans to celebrate idol gods. Rabbis set up the rules to make sure that Jews never drank wine that had been associated with idolatrous offerings, so they boiled it to the point where others wouldn't like it! Flash pasteurization is now more commonly used so that the wine can remain kosher without cooking out its phenolic qualities. Idol worship isn't the problem that it used to be, but in Judaism, tradition is honored and celebrated . . . even in a glass of wine.

17. HOW IS ICE WINE MADE?

Ice wine is a sweet wine made from frozen grapes. I don't mean commercially frozen—I mean frozen on the vine. In -10°, harvesting yield is literally drops of juice or 10% of a regular wine grape cluster. The marble-like grapes are then pressed before going through the fermentation process, which can take three to six months to complete, compared to days or weeks for regular wines. As a result, only small amounts of the sweet nectar are available worldwide, making ice wines relatively expensive. OK, *definitely* expensive.

Although ice wine originated in Austria and Germany, Canada makes most of the market's *Eiswein* in three of its provinces: Quebec, British Columbia, and Ontario. Be watchful of the many fake "iced wines" or "dessert wines" on the market. The grapes must freeze naturally to be legally called ice wine. Although I must admit, some icebox versions are delicious as well: Winemakers use cryoextraction (mechanical freezing) to simulate the effect of a frost, rather than leave the grapes to hang for extended periods as is done with natural ice wines. Fake or real, with their high sugar and acidity levels, these wines can easily age for 30 to 50 years.

18. WHY IS THERE AN INDENTATION IN THE BOTTOM OF A WINE BOTTLE, AND DOES IT MEAN THE WINE IS BETTER?

The indentation is called a punt. Early glassblowers initially made it to make sure the bottle could stand upright without having a sharp edge on the bottom. It isn't needed now. The word stems from the name of a long stick called a puntil rod, or punty, which glassmakers use to gather the molten glass.

The size of the punt doesn't have anything to do with the wine's quality and has no purpose other than marketing to people who think it does.

19. WHY IS THERE A FOIL CAPSULE ON A WINE BOTTLE?

There is a definitive answer to this and it's kinda disgusting. The capsule was placed over the bottle to keep rodents and insects from eating the cork to get to the wine. There were other supposed benefits, one being that it stopped thieves from pulling the cork and replacing the contents. Although capsules are still used to this day, the foil itself came under scrutiny in the 1990s. The original capsules used lead because the alloy is easily malleable and able to be formed into the required shape. When the world became aware of lead poisoning, caps containing lead were banned in 1996 by an amendment to the Federal Food, Drug, and Cosmetic Act and replaced by tin, aluminum, plastic, or wax. And *all* of them are there for purely aesthetic reasons! I still think they are a little gross. Who knows what gunk is underneath those capsules? That's why I recommend wiping off the top of the bottle before pouring.

There are three ways to remove the capsule. The "cute way" uses a foil cutter. I am not a fan of this method, partly because I never seem to have one of them handy. But I also don't like that it cuts under the *first* lip, which can leave the foil too close to the top of the bottle and leads, potentially, to a messy pour. But they obviously work. All you do is push down, squeeze, and turn.

The "fast way" is to simply pull off the foil if it is loose enough. If you can spin the foil around easily, you can probably use this method. This isn't always possible since some capsules are tightly secured to the bottle. One theory (that I disagree with) contends that a bottle with a tight cap indicates that the cork was compromised in some way that made the juice dry and "glued" the capsule tight. It's possible, but it rarely happens.

The third way to remove the capsule is, in my opinion, the correct way. To do it, simply hold the corkscrew knife with your thumb on the back of the blade, cut horizontally halfway around the bottle under the *second* lip. Then flip the knife around and cut the other half. Now give one vertical slice to the top of the bottle. The capsule will peel off. Yes, you can spin it around and around. It's just not as elegant. Hey, do you want to do this correctly or not?

There is another way, of course, which is absolutely incorrect: pulling the cork right through the foil cap. I think the people who do that are the same people who drink milk straight from the carton.

20. HOW IS CLIMATE CHANGE CHANGING WINE?

There was a time when this question could have been "*Is* climate change changing wine?" At this point, however, the answer to that question isn't an opinion or up for debate. Winemakers only need to see and taste their grapes to know that climate change is making a big impact on wine worldwide. In some ways, the wine grape is like the canary in the coal mine for climate change. To be clear, slight changes in climate are pretty much expected in winemaking. That is, after all, why there is a vintage statement on the bottle showing the year the grapes were harvested.

Winemakers check the weather forecast as often as you check your phone. They also keep fastidious records. Harvest dates in Burgundy go back to the 1300s and are among the oldest written records on earth. So, when problems that used to be anomalies become the norm, a winemaker has to step in and make drastic changes, and we see those changes in every country—no matter what its climate. Regions that used to harvest their grapes in mid-October for optimum ripeness levels are having to pick their crop in August! The diurnal shift, the difference between day and night temperature, is crucial to winemaking. Now, due to global warming, the lack of those variances forces the grape through its life cycle too quickly. The skins don't have time to develop properly, which then mutes the acidity and increases sugar levels, resulting in higher alcohol levels. Some vines even shut down completely to protect themselves. Some models have concluded that 2° rise in temperature could shrink some wine regions by 50%.

It is inevitable that some grape varietals won't survive in this new normal. Astoundingly, the heretofore workhorse grape Merlot, responsible for nearly 60% of Bordeaux's wines and grown around the globe, may be one of them! Can you imagine a world without Merlot? Unfathomable! (Peek ahead to question #76 to see how Bordeaux is reacting to climate change.)

One way winemakers can salvage their grapes is by retreating to higher ground. Although not always possible, it is the first option for most vineyards. Other options may include:

- Grow more heat-tolerant varietals with naturally high levels of acidity within fruit.
- Use A.I. sensing technology to assist in vineyard management.
- Consider planting new varietals. Just because particular grapes were originally planted in the region doesn't mean they are the best options now.
- Relax rules to allow new varieties to be grown, particularly hybrids that are more heat tolerant.
- Use modern breeding techniques to research new varieties and rootstocks that can withstand heat and resist disease.

Countries like Poland, Denmark, and Norway, which were not typically known for wine production, are now having weather suitable for growing quality grapes. England is producing world-class sparkling wines. Germany's Spätburgunder (a.k.a. Pinot Noir) is becoming very "New World" in style. And with every change of growing region, the grape also changes, making the wine taste different!

So, Chicken Little didn't get it exactly right, but he was onto something! While the sky isn't falling, the wine map underneath it is definitely shifting.

21. HOW MUCH SUGAR IS IN WINE?

It is hard to state the amount of sugar in wine by varietal or region. There are too many variables that go into the number. To explain, let's first review how the wine process works. During fermentation, the natural or added yeast consumes the natural or added sugar, leaving ethanol (alcohol) as the by-product. When the yeast eats most of the sugars, the wine is considered "dry." But if the yeast is stopped (usually by chilling) before it eats the sugars, it results in a sweeter wine. And so it goes that the more sugars the yeast consumes, the more alcohol is in the wine. Therefore, the higher the alcohol, the lower the sugar. There are exceptions to that rule, however, such as fortified wines, late-harvest/dessert wines, etc.

The actual measurement of sugar in wine is another story. After fermentation, the remaining sugar, called residual sugar, or RS, is either natural to the grape, like fructose and glucose, or is a commercial sweetener, like sucrose, made from sugarcane and beet sugar. A refractometer is used to measure the amount of "gluc/fruc" in wine. The small, handheld tool shows the amount of sugar in the grape, which is called the Brix level. The "degree Brix" is the standard measurement used throughout the food and beverage industry. As it relates to winemaking, one gram of sugar will add about half a gram of alcohol after fermentation.

Unfortunately, nearly everywhere you look, the sugar-to-wine ratio shows grams per liter (g/L), when most wine bottles are 750 milliliters (mL), which is 250 mL less than a liter. So, to determine how much sugar is in your bottle or glass, you'll need a calculator, some math knowledge, and some approximations.

To muddle things even further, there are thousands of grids and charts out there that show the amount of sugar in wines, but there doesn't seem to be a single one that everyone agrees with. My intention is to give you a ballpark range so that you can take your own sweet time when choosing your wine. Here goes:

*4 grams of sugar = 1 teaspoon

Dry wine = 1–4 g/L
Off-dry = 5–15 g/L

Semi-sweet = 25–50 g/L
Sweet = 50–120 g/L
Dessert = 120–220 g/L

Now we can discuss the practice of adding sugar to wine. "Chaptalization," named after Napoleon's agriculture minister Jean-Antoine Chaptal, is the term winemakers use for adding sugar to unfermented juice to balance acidity and, more importantly, raise the final alcohol level. The controversial practice is allowed in some countries and regions and forbidden in others. Champagne is one region where adding sugar is common, even essential, as it is part of *liqueur de tirage*, the mixture that is added to the still wine to force the all-important second fermentation, and the *dosage*, the wine and sugar mix that tops up the bottle and determines the final sweetness level. (See #85 for the full list.)

Mass-produced brands and commercial bulk wines that you can find in every supermarket and wine store around the world use sugar as a corrective measure, so they will have more RS than premium wines that are made with less intervention. This explains why some people claim they get a headache when they drink cheap wines. My answer to this problem is simple: Drink better wine!

Tale from the Wine Floor

Customer: Is this wine sweet?

Me: Yes.

C: I mean really sweet. My wife only drinks wine if it's *really* sweet.

M: Well, not knowing your wife and her perception of sweetness, I can only tell you that it is definitely a sweet wine.

C: I guess it's OK. She'll only add sugar to it anyway.

M: Wait . . . your wife adds sugar to her wine?

C: All the time.

M: Maybe she just doesn't like wine.

C: She doesn't, but her doctor told her it's good for her.

Chapter Two

TASTING

22. WHAT IS THE BEST WAY TO LEARN ABOUT WINE?

The best way to learn about wine is to drink it! Taste any wine you can get your hands on, and take notes. There is no need to use "wine speak" in your descriptions or to write in a fancy journal. A plain old notebook and language you understand are all that matters. A good starting point is to taste a fruit-forward wine with few tannins, like a Gamay (the grape in Beaujolais). Describe the fruit like you would describe any fruit. Is it ripe enough for you? How sweet is it? Write it down. (I like to keep it simple and use general terms like low, medium, medium+, and high.) How much acidity is there? Does it remind you of any other fruits? Write it down. How is the mouthfeel or weight? Write it down. Are there any spices that come to mind? Do you like it?!

Now try the same grape from a different country before moving on. For example, Malbec, although a signature grape of Argentina, is originally grown in Cahors, France. An example using a white wine is Sauvignon Blanc. The grape, when it's from Marlborough, New Zealand, can have a totally different taste profile when it's grown in France's Sancerre region, which is also different from when it's grown in Napa, California.

Some grapes may even have a different name in another country. It's easy to look up, i.e., *Hey Siri, what is a synonym for the wine grape Pinot Noir?*

Next, taste it with and without food. To make it more fun (and less expensive!), put together a group of friends. (See #56, "How do I host a wine tasting at home?") Lastly, give it your score! Don't forget to add your initials after it like you're a big shot.

Fun fact: A wine lover is called a "oenophile." The word is a combination of two Greek words, *oinos* (wine) and *phile* (lover of). The French put the two together to create the word *oenophile* in the mid-1800s. It does not mean connoisseur, which is a person with expert knowledge or training in the fine arts.

🍷ale from the Wine Floor

Customer: Eh, are you the wine guy here?

Me: Yes, how can I help you today?

C: I need you to pick out a bottle of wine for a gift.

M: I'd be happy to. Would you like a certain varietal or style of wine?

C: I dunno, just pick one.

M: OK. Would you like a dry or sweet wine?

C: I dunno. It don't matter. Just pick one.

M: How much did you want to spend?

C: I don't care, I'm good. Just pick one.

M: Got it. This wine here is a blend. I'm choosing this because . . .

C: Stop. I'll take it.

M: I just thought I'd tell you a little about it in case someone asks.

C: Nah, I don't wanna know.

M: Now you have me curious. Why not?

C: 'Cause it's like this: If I know and my wife asks, "Why'd you get this one?" I'll hafta go into this big thing and all. This way, I can say, "The wine guy picked it! Whaddaya want from me?!"

M: This one.

C: Thanks.

23. WHAT'S THE DIFFERENCE BETWEEN SWEET WINE AND DRY WINE?

This question is not as straightforward as it seems. For starters, when it comes to wine, the opposite of sweet is dry, not bitter. Plus, most wine drinkers mix up the words *fruity* and *sweet*. The absence of sweetness does not mean the absence of fruit. On top of that, individual perceptions of sweetness are affected by a person's age, weight, whether they smoke or not, and their cultural upbringing. So, what you think is sweet or dry may not match the bottle's labeling of sweet or dry, if it's on the label at all!

From the winemaker's perspective, the label/description of sweetness of wine has nothing to do with the subjective opinions I just described but is determined by how much residual sugar (RS) is in the wine.

As previously mentioned, the way a wine is processed determines its RS. Remember that grapes have natural sugar and yeast. During fermentation the yeast eats the sugar, and alcohol is the by-product. However, to make a sweet wine, a winemaker stops the sugar from turning into alcohol, leaving RS in the wine. They could also pick the grapes later (actually called late harvest) when they are more fully ripe and, therefore, have more natural sugar.

There are a few other ways winemakers can make sweet wine:

- They can let the grapes dry in the sun, which will concentrate the sugars.
- They can let the grapes freeze on the vine before picking and pressing them, as is done to make ice wine. The frozen water stays with the grape, leaving just the condensed nectar to be fermented.
- They can also let the grapes rot! *Botrytis* is the term but is oddly known as "Noble Rot." It intensifies the sweetness level and adds complexity to the wine.
- They can just add sugar.

To make things even more confusing, winemakers use vague terms like "semi-sweet" and "off-dry." Again, both terms refer to the RS remaining in the wine and not a wine professional's opinion.

With all of this in mind, making a semi-sweet wine recommendation for a customer can be as hard as guessing their favorite color. Making it harder still, people seem to talk dry but drink sweet.

So, there's the answer, short but sweet.

Tale from the Wine Floor

Customer: Can you tell me if this wine is sweet?

Me: Yes, that is sweet. As a matter of fact, everything in this aisle is sweet.

C: So, this wine is sweet too?

M: Yep, that's sweet as well. ALL of these are sweet. This is our sweet aisle.

C: So, this one?

M: Sweet. That's why we put it in this aisle.

C: And this?

M: Yes. Once again, EVERY bottle of wine in this aisle is, in fact, sweet.

C: How 'bout this one?

M: Let me see if I can make it simpler for you. ALL of these bottles, from here to here, are sweet wines. We put all of our sweet wines together in one aisle and called it our sweet aisle. So, if the bottle is in this aisle, it's a sweet wine. Does that help?

C: So, where are all of your dry wines?

24. WHAT'S THE DIFFERENCE BETWEEN AROMA AND BOUQUET?

The words *aroma* and *bouquet* are often used interchangeably, but they actually have different meanings. The "aroma" of wine comes from the grape itself. So, if you are using words that describe fruit, like raspberry, strawberry, lemon, black currant, etc., it's the aroma. Wine geeks call them *primary* notes or flavors.

On the other hand, the bouquet (like a bouquet of flowers) is derived not just from the individual components (the grapes) but from the entire winemaking and fermenting process—the sum of its parts. Explanations of "bouquet" use non-fruit descriptors like toast, nuts, leather, mocha, vanilla, cedar, etc. Logically, bouquet descriptors are called *secondary* notes. But once again, wine geeks need a geekier word, so they call them "tertiary" flavors. I never use the term myself because I'd only have to say "tertiary" means "secondary." Why not just call them that in the *primary* place?

25. WHY IS RED WINE CONSIDERED BETTER FOR YOU THAN WHITE WINE?

Red wine has long been thought of as the healthier choice due to the antioxidant *resveratrol* found in red grape skins. The compound is also in the skins of white grapes but generally in much lower amounts. Remarkably, there are 1,600 other antioxidant and polyphenol compounds in grapes that are known to have health benefits, and yet every article on wine is filled with the word "may." Wine *may* reduce heart disease, *may* reduce the risk of stroke and diabetes, *may* reduce inflammation, etc. More recent studies are finding that alcohol itself *may* be the primary reason for the health claims of wine consumption. It can relax blood vessels and increase levels of HDL, the "good" cholesterol.

It should go without saying, but I'll say it anyway: Moderation is the more important thing to remember as there is no study that advocates drinking more than one or two glasses of wine a day—a fact I forget myself. Often.

Below is a table showing nutrient contents for five ounces of both red and white wine. As defined by the Food and Drug Administration (FDA), the Recommended Daily Intake (RDI) defines how much of a nutrient a person should have based on their age and gender. But the RDI is different from the Daily Value (DV), which is the overall number, regardless of age and gender. However, the DV tends to be higher than the RDI, so an Upper Limit (UL) is required to let you know the highest amount you should consume. Confused? So am I (SAI).

	Red Wine	White Wine
Calories	125	121
Carbs	4 grams	4 grams
Sugars	1 gram	1 gram
Manganese	10% of the RDI	9% of the RDI
Potassium	5% of the RDI	3% of the RDI
Magnesium	4% of the RDI	4% of the RDI
Vitamin B6	4% of the RDI	4% of the RDI
Iron	4% of the RDI	2% of the RDI
Riboflavin	3% of the RDI	1% of the RDI
Phosphorus	3% of the RDI	3% of the RDI
Niacin	2% of the RDI	1% of the RDI
Calcium, vitamin K, zinc	1% of the RDI	1% of the RDI

Tale from the Wine Floor

Customer: Can you recommend a wine that's good for gout?

Me: I have absolutely no clue. What makes you think there's a wine for gout?

C: 'Cause they say wine is good for you?

M: They do . . . in moderation.

C: OK, so I have gout and I want a wine that's good for that?

26. HOW CAN I EXPAND MY WINE PALATE?

Palate (as it is used here) is defined as a person's appreciation of taste and flavor, which is always preceded by the pretentious terms "sophisticated" or "discriminating." The question is this: Is your palate affected by nature or nurture? The answer is that it is affected by both.

In the "nature" camp are the taste buds you inherited. Each of us is born with approximately 10,000 taste buds that regenerate every 10 days or so. The number of taste buds varies from person to person. Approximately 25% of the population is thought to be supertasters. A supertaster experiences taste with far greater intensity and can discern distinct flavors more than the average person. Interestingly, 35% of women and only 15% of men have above-average taste buds. Ethnicity also plays a role. Asians seem to have a higher proportion of supertasters. Age also affects our ability to taste. As we grow older, the regeneration of taste buds slows down, taking a nosedive at about age 60. (Kinda like your hair now that I think about it.) So, you can't actually change the number of taste buds you possess.

On the other hand, nurture plays an important role in your taste perceptions and preferences. Basically, this is how it works: When you taste/smell something (and yes, your sense of smell plays a big part in taste), your brain searches to associate that taste/smell with a time you've had it before. You'll recognize this, for example, when you smell something that immediately takes you back to a different time or place. For me, the smell of spaghetti gravy (yes, I said gravy, not sauce; it's a Philly thing) simmering on the stove transports me right back to my childhood kitchen. If there is no record in your memory of that particular taste/smell, your brain will make a new record for it. So, as you taste new and different flavors, you are growing the number of reference points in your brain, in essence "expanding" your palate.

But of course, it's more complicated than that. People are more comfortable with familiar flavors, and familiar flavors are the result of culture and upbringing. For example, my mother didn't like or cook with cumin, so I don't like cumin. (The same goes for cilantro, but that's a whole different issue.)

If you want to expand or widen your wine palate, you can do it. It just requires focus. Breaking old habits, as you may know, is more of a mindfulness practice than a physical change. When I teach wine, I remind the class to use all of their senses. For instance, I often encounter people who drink only sweet wine. Before tasting the first wine without residual sugar, I ask them to focus on the fruit. What kind of fruit do you taste? Is it blackberries, cherries, lemons, or green apples? Telling them up front that they will not notice overt sweetness saves them from the sucker punch they will inevitably feel with that first sip. Think of this analogy: If you were blindfolded and told you were going to drink orange juice but instead were given cranberry juice, you would think it was the worst orange juice you ever tasted. In fact, there was nothing wrong with either juice—you just weren't expecting that taste.

As you drink more styles of wine, you will recognize and become more familiar with more flavors, which will then be easier to discern and describe, since you will now have more reference points in your internal filing system.

Tale from the Wine Floor

Customer: Can you recommend a really good wine?

Me: Sure. Give me a price range.

C: Well, I heard cheap wines are better than expensive wines.

M: Who told you that?

C: My brother-in-law, and he *knows* wine!

M: Well, I could show you a good cheap wine and a bad expensive wine, but I can't agree with your brother-in-law that cheap wines are better than expensive wines.

C: OK, then just show me that good cheap one.

M: Who is it for?

C: My brother-in-law.

27. WHAT ARE "LEGS" IN A WINEGLASS?

"Legs" in a wineglass are the tears that stream down the side of the glass after you swirl it. This scientific phenomenon is the result of fluid surface tension caused by the evaporation of alcohol. It is known as the Gibbs-Marangoni effect, named after the two guys who discovered it.

Wine is a mixture of alcohol and water. Thicker and slower legs indicate a higher level of alcohol content in the wine. Interestingly, while sugar content determines how much alcohol a wine will have, it does not affect the number or size of the legs you see. It does, however, make them drop slower. Whatever the speed and shape they take, the legs tell you nothing about the wine's quality, only the alcohol.

An exception is dessert wines. The residual sugar (RS) changes the viscosity of the wine, which affects the legs.

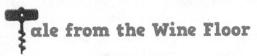

Tale from the Wine Floor

Female model to pour for store tasting:

Model: Hello, are you Jimmy?

Me: Yes.

Model: I'm here to pour xxxxx.

Me: OK. Let me get you set up right here.

(15 minutes later)

Me: I'm always curious, how'd you get this job?

Model: I have nice legs.

Me: Uh, no . . . I meant . . . never mind.

28. WHAT ARE HORIZONTAL AND VERTICAL TASTINGS?

"Horizontal" and "vertical" are both terms to describe ways of tasting wines side by side. The basic premise for both is that by isolating the year or producer, you can get a sense of how either can influence what's in the bottle.

In a horizontal tasting, multiple wines from the same year are sampled. They can feature a range of wines from a particular area, as long as they are from the same vintage. Horizontal tastings are a great educational tool for wine newbies to get a basic understanding of the wine regions of the world.

In a vertical tasting the same wine from multiple years is sampled. Since you are tasting a singular wine, vertical tastings help producers better understand how their wines evolve over time. Wineries sometimes organize vertical tastings of their library wines to see if their wines are aging as gracefully as they had hoped. Serious collectors love hosting vertical tastings to see how their prized wines are developing . . . and to show off their collections.

Side note: I am always available for anyone showing off their wine collection. What can I say? I'm a giver.

↑ale from the Wine Floor

Customer: This is Mondavi, right?

Me: Right.

C: He makes the best wine, right?

M: Mondavi makes some nice wine, yeah.

C: But not the best?

M: Well, wine is so subjective it's impossible to say.

C: It's what?

M: What?

C: That word.

M: You mean "subjective"?

C: Yeah, that.

M: Oh, it means that everybody's taste is different. It's in the eye of the beholder.

C: What is?

M: Wine. That phrase is talking about beauty, but wine is like that too.

C: So, wine is like beauty?

M: Exactly!

C: So, does Mondavi make any like that?

M: Like what?

C: Subjective ones.

29. WHAT'S THE POINT OF BLIND TASTINGS?

When I say "blind tastings," I am not talking about closing your eyes and drinking. Blind tasting refers to tasting a wine with no prior knowledge as to its identity. Marketers do the exact same thing by comparing their product to competitors (remember the Pepsi Challenge?). It can be performed with wine, as well as any other beverage, but when a professional does it, it's called "deductive tasting." With a focus on the sight, smells, and flavors of the liquid, the tasters attempt to deduce not only the grape variety but also the wine's geographical home and quality level. I assure you it isn't a parlor trick. It involves hours of study and practice to be able to figure out what the wine is by figuring out what it isn't. Every wine has key indicators that will help you find its identity. If your line of work is to purchase and recommend wines (as mine is), blind tastings are a crucial part of judging a wine's quality.

Wine distributors are in the business of sales and will naturally try to glorify the product they are selling. But only by knowing the traits of each wine can you break it down and impartially decide on its quality level. Deductive tasting provides the necessary tools not only to determine whether it's worth the price but also how to relay those traits to customers (which is more difficult than you would imagine). If you have a mental database of wines you've smelled, tasted, and analyzed deductively, it's much easier to recommend a wine the customer will love. For the average wine drinker, blind tastings are actually a lot of fun. Trying it with wine-loving friends is something I definitely recommend! (For a how-to, go to question #56.)

A surprising benefit that I've found from "blinding wine" comes from being wrong. I distinctly remember one blind tasting where I concluded the wine was a Pinot Noir. I was sure of it. It had notes of cherry and strawberry, was low in tannins and high in acidity—all the traits of a New World Pinot. It turned out to be a Barbera: a fruit-forward wine to be sure, but from Italy's Piedmont region—certainly not a New World wine. I consoled myself by remembering that it's helpful to know which grapes are similar to other grapes so that I can offer an alternative selection for an out-of-stock wine or a recommendation that is in the same wheelhouse of a customer's preference. For example, a fresh, fruity, and light Beaujolais from France can be quite similar to a Dolcetto from Italy. If a customer says they like Beaujolais, Dolcetto would be the perfect recommendation. If I suggest, say, a dry and tannic, full-bodied wine, like Aglianico, my credibility will most certainly come to an abrupt end once the customer tastes my recommendation.

Without question, the most famous and game-changing blind tasting was the Paris Wine Tasting of May 24, 1976. The first of its kind, the event became known as the Judgement of Paris. It pitted French wines from Burgundy and Bordeaux against their American counterparts. The results of that tasting, which America won in both categories, put Napa, California, and American wines in general on the world stage, with investors and buyers from around the globe clamoring for wines from the New World. The tasting, immortalized in the movie *Bottle Shock*, cannot be overestimated in the story of American wine. The winning bottles are, after all, in Washington, DC's Smithsonian National Museum of American History!

So, try blinding wines. I think you'll find it to be a real eye-opener.

30. WHAT'S THE DIFFERENCE BETWEEN OLD WORLD AND NEW WORLD WINES?

Old World means Europe, and New World means everything else. But the differences go to the very heart of wine. Where and how the grape is grown contributes to what the French call *terroir*, or sense of place. In practical terms, New World wines from the Southern Hemisphere and the United States are generally more "fruit-forward" (pretentious somm-speak meaning fruitier) and, to some, fresher tasting. On the other hand, Old World wines from Europe are known for their elegance, and they are considered better with food. But the distinctions go much further. Here are some of the contrasts between Old and New World wines:

Old World	New World
Names the place	Names the grape
Tradition	Innovation
Old-fashioned ways are preserved	Technology is revered
Fixed growing regions	Flexible growing regions
Winemaking as an art	Winemaking as a science
Credit to the vineyard	Credit to the winemaker

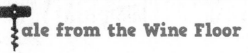

ale from the Wine Floor

Customer (older lady): What the heck are we listening to?

Me: I'm not sure . . . just a pop station on the radio.

C: Can't you put on real music?

M: Well, what do you mean by "real music"?

C: Sinatra!

M: Let me see what I can do.

(I change the station and a Frank Sinatra song comes on.)

NEW customer (young guy): What the f*** are we listening to?

M: That's Sinatra.

C: Never heard of them.

31. ARE ROSÉ AND BLUSH WINES THE SAME?

Rosé, pink, or blush wines can be made using any red grape varietal and can be sweet, dry, or sparkling. Although the ancient Greeks made the first Rosé wines, most Americans were introduced to the style from two Portuguese brands: Mateus and Lancers. However, it wasn't until 1972 that the Rosé wine world changed forever as a result of winemaker Bob Trinchero's happy accident. His Zinfandel grape juice went through a stuck fermentation in which the yeast stalled before the sugar could be converted into alcohol. Expecting to throw out the lot, he was pleasantly surprised by the soft and sweet wine he had inadvertently invented. From that point on, Sutter Home became synonymous with White Zinfandel.

The term "blush wine" has been used on labels in the United States for decades for everything from White Zinfandel and White Merlot to Pink Moscato. At this point, the terms are interchangeable. There are three principal methods by which Rosé is made:

- **Skin contact:** The juice is left in contact with the red skins for a short period of time before being removed. The pink-stained juice is then left to ferment as usual.
- **Saignée:** This method bleeds off some of the red wine from a barrel before the maceration process makes the juice darker.
- **Blending:** Although only sanctioned in the Champagne region, this method simply blends red and white wine together.

Tale from the Wine Floor

Customer: Are you the wine guy here?

Me: Yes.

C: I'm looking for the Zinfandels.

M: White Zin or red?

C: There is white and red?

M: Yes. Here is the white.

C: That's pink! Why don't they call it pink Zinfandel?

M: Sounds logical enough . . . but they don't.

C: So, this White Merlot is the same?

M: Yup.

C: How 'bout these pink wines here?

M: They are actually Rosé wines. They're made with different grapes and a little skin contact that makes them pink.

C: What's the difference between Rosé wines and pink wines?

M: Nothin'.

C: Well, how do you know what grape it is if it just says Rosé?

M: I usually read the back label.

C: And they pay you for this job?

32. WHY DOES THE SAME WINE SOMETIMES TASTE DIFFERENT?

First and foremost, *you* changed. To be more specific, your taste buds have. There are three different types of taste buds—50 to 100 cells each—that play different roles in sensing the five primary tastes: sweet, sour, salt, bitter, and the newly recognized but hard to define umami.

With an average lifespan of just 10 days, a new generation of papillae (taste buds) may not have the same likes and dislikes as the old generation. I'm sure you can name things you used to love as a child, but you wouldn't touch now, and vice versa. To my point: When did you start liking asparagus? Cigarette smoking and weight change are also factors in your new taste buds' size and shape, and aging itself causes some not to regenerate at all. Researchers believe that the loss of taste starts to kick in around 60 years old. Your sense of smell also plays a starring role in taste, which also deteriorates with age. Aren't you sorry you asked?

Let's get back to the wine. Storage, transportation, glassware, and serving temperature are all factors that make that familiar bottle slightly unfamiliar. However, the biggest factor is time. The wine is aging just like you. It's the circle of life. Hakuna Matata!

33. WHAT ARE THE BEST "EASY DRINKING" WINES?

That's a fair question. The term "easy to drink" shouldn't be an insult to a wine. Yet you rarely, if ever, see those words on a store shelf talker or restaurant wine list. Retailers somehow equate an uncomplicated wine as unsophisticated. As a result, we have wine experts and sommeliers forced to use arcane (and sometimes insane) words to describe a given bottle. In the documentary *SOMM*, Ian Cauble described a wine's smell as "cut garden hose and freshly opened can of tennis balls." That doesn't sound like easy drinking, does it?

So, what does "easy to drink" mean, exactly? To my mind, it's a wine that doesn't need to be decanted, doesn't need to be swirled in a special glass, and doesn't need to be dissected or even described. It won't tire out your palate with high alcohol, high acid, or high *anything*, yet it won't be a fruit bomb. And it certainly doesn't need to pair with wild boar! Easy drinkers pair best with Tuesday-night tacos or Friday-night pizza. They should be inexpensive, but that doesn't mean cheap. Let's call it good-bang-for-the-buck wine.

Possibly the best example of an easy-drinking wine that got lost in a difficult wine world is Merlot. One line from a Hollywood actor in the movie *Sideways*, and it became the most uncool of varietals. What people like about Merlot (at least prior to the movie) is its softer-than-Cab character. Yet here we are, almost 20 years after Paul Giamatti uttered the line, "If anyone orders Merlot, I'm leaving. I am *not* drinking any [expletive] Merlot!" and the grape hasn't fully recovered. Ironically, Giamatti's character, Miles Raymond, hated Merlot *because* it was so easy to drink.

Maybe wine scores are the problem. Critics must fall into the "bigger is better" trap after tasting 50 wines at a shot. I can't imagine a high-scoring wine being described as "easy." Don't get me wrong, I love geeking out about a first-growth Bordeaux that I was saving for a special occasion. But I wouldn't put that much attention (or money) into a wine if the occasion was burgers and hot dogs on a lazy summer Saturday.

Perhaps the quintessential easy drinker is Rosé. For the first time, in 2019, Rosé sales surpassed Prosecco. It is quickly becoming America's favorite wine. The heretofore summer-only varietal is inching its way into a year-round category, mainly due to the container—wine in cans! What's more "easy drinking" than that?

Generally speaking, white wines are easier to drink than reds, though not the over-oaked Chardonnays or overly herbal Sauvignon Blancs. "Herbal" isn't a word in the easy-drinking dictionary.

My gauge for an easy-drinking wine is what I call my "empty bottle" test. If I grab the bottle for another glass only to find we finished it already, that's an easy-drinking bottle of wine. Though with that criteria, shouldn't all wine be easy drinking?

Here is my shortlist of easy-drinking wines. There is no geography lesson required, no tasting notes, and no special glass—just simple and delicious wines that should be more popular.

White:
- Albariño
- Torrontés
- Garganega
- Vermentino
- Vinho Verde

Red:
- Beaujolais
- Dolcetto
- Barbera
- Montepulciano d'Abruzzo
- Grenache

ale from the Wine Floor

Customer (husband): Can you help us?

Customer (wife): We don't need any help.

Customer (husband): Yes, we do. Where are your Pinot Grigios?

Customer (wife): But I don't want Pinot Grigio!

Customer (husband): I thought you said you liked Pinot Grigio?

Customer (wife): I do, but I don't want that. I want a White Zinfandel.

Customer (husband): But I don't want White Zinfandel!

Customer (wife): What do you want?

Customer (husband): Pinot Grigio!

Me: Um . . . the Pinot Grigios are here, and the White Zins are over there.

Customer (wife): If we need you, where will you be?

Me: At lunch.

Chapter Three
BUYING

Tale from the Wine Floor

Customer: Can you help me pick out a wine for a Christmas present?

Me: Absolutely. Who is it for?

C: My wife, but it has to have the right name on it.

M: Got it. Here's a semi-sweet wine from Italy called Mi Amore ("my love"). How's that?

C: What else?

M: Alright. Here's a Moscato called Mommy's Time Out.

C: What else?

M: How 'bout a wine called Santa's Little Helper?

C: Nah.

M: Little Black Dress? Fifty Shades of Grey? Skinny Girl? Love Noir?

C: Nope.

M: Well, that's all I can think of.

C: Wait! There! That one?

M: Sweet Bitch?

C: That's it!

M: Umm . . . Merry Christmas.

34. WHY ARE SOME WINES SO EXPENSIVE?

You wouldn't think it at first, but wine is a commodity. And like all market-driven commodities, prices are based on supply and demand. There are, however, many factors that go into making a wine good enough to command a price no matter how high or low.

The first factor is where the grapes are grown. Trial and error over thousands of years have taught us where certain grapes grow best. Basically, a wine that simply states "California" as the appellation (think: hometown) is less expensive than one from a specific vineyard (think: your home address). That is because the "California" wine can be made using grapes grown anywhere in the entire state. Wine from more specific regions may cost more because the land itself costs more. For example, an acre of land in Napa can cost a million dollars or more, and that's before a single grape is grown. More specific regions also have limited availability. The less available product there is, the more it costs.

The next factor is the type of oak used to create the barrels and the amount of time the juice spends in it. That can add two to four dollars to the price of the bottle. Then there is the bottle itself, which can add one to three dollars. After that, add in the cost of labor, the cork, and the packaging.

Another factor in determining the price for a bottle of wine is what score or rating it may have gotten. Prices do fluctuate based on who gave it that score. The score may not matter to the average customer, but it can impact the price on the shelf.

Obviously, a key factor is the taste of the wine. If the wine doesn't taste good, consumers rightfully think it isn't worth the price no matter what the number. On the other hand, consumers are willing to pay a higher price for a delicious wine.

And finally, *when* the wine was made can have an impact on its price. Wine technology has made it easier to fix problems in the vineyard and in the cellar. The generation that stood firm on traditional winemaking has given way to their children, who have a more modern approach to winemaking and cost-saving techniques.

In my opinion, a $20 bottle today is better than a bottle costing three times that price just 20 years ago. In the future, it will become more expensive as winemakers take measures to battle climate change (see question #20, "How is climate change changing wine?"), but as of now, great wines can be had at incredibly reasonable prices.

Fun fact: The most expensive bottles of wine ever sold were two bottles of 1945 Romanée-Conti, fetching $558,000 and $496,000 in 2018.

Tale from the Wine Floor

Customer: Why would anyone pay 50 bucks or more for a bottle of wine?

Me: Well, sometimes it's supply and demand. There is just less of that high-priced wine, making it rare . . . and people pay for "rare."

C: Not me.

M: You do, actually, just maybe not for wine.

C: Nope, I don't pay for rare anything.

M: I guess then you're the anomaly. By the way, that's a nice ring you have on. Cubic zirconia?

C: No, this is a real diamond.

M: Why would anybody pay for a diamond when cubic zirconia looks the same?

35. IS WINE VEGAN?

You would think that all wines are vegan. They're grapes, for goodness' sake! Well, you'd be wrong. They *could* be vegan, but winemakers sometimes use a clarifying process to prevent the juice from being hazy and unappealing. That clarifying process, called "fining," traditionally uses casein (a milk protein), albumin (egg whites), gelatin (animal protein), and isinglass (fish bladder protein) to stabilize the wine. The fining agent attracts and coagulates the molecules—such as proteins and tartrates—making them large enough to be filtered out while it's still in the cellar. The resulting wine is clear and bright, but now it is *not* vegan. And, by the way, organic wine doesn't guarantee that it's vegan either.

If a winemaker chooses to make a vegan wine, there are other fining agents that could be used. Bentonite, for instance, is a clay created by volcanic ash. It is effective in absorbing the particles without adversely affecting the taste. However, increasing numbers of natural winemakers choose to leave the wines in their natural hazy state.

36. HOW DO I KNOW IF A WINE IS GLUTEN-FREE?

All wine is naturally gluten-free. Yes, winemakers may use fining agents that add gluten to remove unwanted elements of the wine. And yes, the gluten remains behind as sediment at the bottom of the storage container in the cellar when the wine is filtered and transferred to bottles. But after fining, any remaining traces of gluten falls below 20 parts per million, which is the limit set by the Food and Drug Administration (FDA) for labeling items gluten-free. So even after using this type of fining process, a bottle may still legally be labeled "gluten-free."

Tale from the Wine Floor

Customer: Why is this French wine so expensive?

M: Actually, because of Napoleon.

C: The pastry?

M: No, the emperor.

37. HOW DO I ORDER WINE AT A RESTAURANT?

Whether it's a single page or a leather-bound book the size of *War and Peace*, the wine list at a restaurant shouldn't be intimidating at all. As long as you have a little background knowledge, you'll be fine. So just swallow your fear in anticipation of swallowing some delicious wine. The first thing you'll need to know is that each bottle has five five-ounce glasses in it and should serve three people (OK, not if *I* am one of those three, but I digress. . . .). Before you order, ask who is drinking wine. If one person is more enthusiastic about it than the others, ask if they would like to pick a wine for the table—unless you want to, in which case, don't ask, just go for it!

Check out how the list is organized. Some are by country, some are by varietal (that's the name of the grape), and some are categorized by the body of the wine (light, medium, or full). Ask if anyone has favorites. If it's just you and your date, you might ask what he or she is ordering for dinner. If you're with a group, it's probably too early to know, in which case, a sparkling wine makes a nice introduction to the meal. Otherwise, you can hold off ordering your wine until everyone orders their food.

If you are at a fancy-schmancy place, there could be a sommelier—yay! They are not the stodgy wine snobs that you think they are. In fact, they want nothing more than to help you pick a wine that you'll love, at the price you want to spend, so that you'll come back. If you really don't know too much about wine or what wine you like, it's better to tell them that than to pretend you're a connoisseur. If you do have an idea of what you want, tell them what you're thinking and possibly what you're ordering for your meal. Ask, "Do you have any recommendations?" and tell them what you'd like to spend. Talking about the price you'd like to spend doesn't have to be awkward. You can use a general term like "mid-range" or even be more specific like "under $60." That price may not be your idea of "mid-range," but a restaurant typically charges three times the bottle's wholesale cost, with cheaper bottles marked up higher than expensive ones. Therefore a $20 bottle on a store shelf could be $60 on a restaurant menu. A restaurant will also charge full mark-up for popular brand names because they sell no matter what. So, a night out to dinner may be the perfect time to try something new that has a better price tag. My favorite technique is to ask the somm what he or she has tried lately that might be "under the radar." Those wines, like an under-appreciated grape from an obscure region, can be great and have a great price. Somms love recommending the odd ones.

OK, so now you've decided. They bring out the bottle and show it to you. A quick check that it's the wine *and vintage* you chose is all that's needed. Give a nod and continue enjoying your guests. They will then open the bottle and present you with the cork. Don't make a big deal of it by waving it under your nose for a week; just squeeze it to make sure it doesn't crumble in your hand. That would tell you the wine was badly stored. It still might be OK even if it is slightly crumbly, but it's worth noting. You will then be poured a small sample for you to taste. Again, don't make a big deal of it. You're not scoring the wine for *Wine Spectator*. Give a taste, give another nod, and say, "That's great, thank you." (By the way, I like to pour the wine at the table, so I'll tell them that.)

How 'bout if the wine *is* bad? "Bad" does not mean you ordered a wine that you didn't like. It means the wine smells and tastes like wet cardboard or the cork . . . or like nothing! This is rare—so rare that it might happen only a few times in your life. But if it does, kindly ask the somm or server to take a taste. They will agree if it's bad (they may even humor you and agree that the wine is bad even if it's not). They will then offer you another bottle, or you could choose your second choice—the one you almost ordered.

Wine *does* make the meal. Don't deprive yourself of the pleasure of sharing a nice bottle due to a preconceived notion that you might do or say something wrong. You won't. You can't.

38. DO WINE SCORES REALLY MATTER?

A wine "score" is the grade given to a bottle of wine by wine critics. Whether they really matter depends on whom you ask. Scores matter to trade magazines and retailers because they help to sell wine. Scores matter to a winemaker because a high score naturally allows for a supply-and-demand price increase. Millennials and Gen-Xers show less reliance on scores in general—though, are Facebook "likes" or shout-outs from Instagram influencers any different?

There isn't a day that goes by that I don't get asked about or see a wine score. Given the thousands of bottles available for purchase, it is the quickest way to get an opinion about the wine's quality. But whose opinion is it? And why would you trust the opinion of someone you never met or never heard of? Here's my analogy: If your brother-in-law recommended a movie that you then saw and hated, odds are you wouldn't ask for your brother-in-law's opinion on movies again. The same goes for wine critics.

So how did we get to this point in wine? There were undoubtedly "experts" who critiqued wines before 1982, but that was the year the wine-scoring changed, and in turn, the world of wine changed. This is what happened.

One writer from a small town in Maryland said the 1982 vintage in Bordeaux was "stunning" and urged the readers of his self-published newsletter to buy the wine. It wasn't just that his opinion differed from others; it was how he *scored* the wine that was so influential. Robert McDowell Parker Jr. was visiting his high school sweetheart at the University of Strasbourg when he had his wine epiphany. Returning home, he wrote his opinions of the wines he had tasted and published them in a newsletter for his friends. Taking his cue from Ralph Nader's popular magazine *The Consumer Advocate*, he called his modest effort *The Wine Advocate*. Instead of giving one to five stars as traditionally seen with movies or the 20-point scale that was used in Britain, he used the 100-point marking system that was understood by every American student. He had no idea, however, that anything under 90 would be considered subpar, but that became the benchmark. He also thought a wine receiving a score in the 80-to-85 range was perfectly fine. I dare say you will never see a shelf talker touting an 82 score today! Never was the insanity of scoring vs. sales more evident than in the case of Two Buck Chuck, where a bottle of Charles Shaw wine, which cost $1.99, was given a high score and went on to sell as if it were a first-growth Bordeaux.

Now, I'm not against wine scores—they definitely serve a purpose. But I think you should take note of who gave the score and whether their palate matches your own after having tried a few of their recommendations. Whether you use *Wine Spectator*, *Wine Advocate*, or *Wine Enthusiast* for a score (or even *Wine-Searcher*, which calculates the average), relying solely on anyone's palate other than your own is risky and may deprive you of exploring a grape or region that you might love—even if it didn't get a 90+ score. But that's just my opinion.

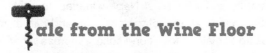

ale from the Wine Floor

Customer: Is this a good wine?

Me: It is. In fact, we just got that one in. It's a blend of . . .

C: What score did it get?

M: I'm not sure off the top of my head, but I just tasted this with the staff, and we all would easily give it a 90.

C: But I only buy wines that actually *got* a 90.

M: From who?

C: It doesn't matter. As long as it got a 90.

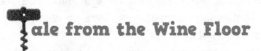

ale from the Wine Floor

Customer: Where are your cheap wines?

Me: We have a whole section of what we call "close-out" wines. These are all pretty cheap.

C: I mean *real* cheap!

M: Got it. This wine is only $3.99.

C: What score did it get?

M: What score did it get? It's $3.99!

39. DO VINTAGES REALLY MATTER?

A vintage is the year in which grapes were harvested. It does not mean the year it comes to your local wine shop. Some wineries hold back wine in barrels or bottles until their work of art is ready. On the other hand, commercial wineries that produce wine in large quantities are more concerned with consistency. They can (and do) manipulate the juice from any vintage to make it taste virtually the same from year to year.

Winemaking aside, vintages are more or less an issue depending on geography and wine laws. Napa wineries, for instance, have fairly consistent temperatures to contend with and can plant grapes of their choosing to match those conditions. Whereas wineries in, say, Bordeaux and Burgundy, which have borderline ripening conditions and specific traditional grapes, seem to be in a constant tug-of-war with Mother Nature. Modern wine techniques and innovations have helped in that regard, but countries like France and Italy produce wines where vintages are a constant fascination with collectors. And yet, most wine drinkers haven't the slightest interest in or care about the vintage. You should know that wine stores stock the most current vintage. If it does have a few years of age, it's probably because it hasn't sold. In fact, it's a dead giveaway that a customer doesn't know what they're talking about when they ask for a specific vintage of an everyday "value" wine that was made to taste the same year after year.

A restaurant wine list is a different story. Most of the bottles on their menu will most certainly list the vintage. I still would recommend the most current vintage because you have no idea how and where the wines have been stored.

Collectors and connoisseurs are on a constant quest for the best wines from the highest-rated years and memorize *Wine Spectator*'s Vintage Chart, but that is a fairly small percentage of wine drinkers.

So, if you're wondering if vintages really matter, the answer is yes and no. Yes, it matters for some wines (and some people), and no, it doesn't matter for others. How's that for a crystal-clear answer?

ale from the Wine Floor

Customer: What does "vintage" actually mean?

Me: That's the year the grapes were grown.

C: Why don't wines give an expiration date like milk?

M: Well, there's no way to know when that is. Wine ages differently depending on how it's stored.

C: Like milk!

M: I guess.

C: They could put them in plastic too . . . like milk!

M: Uh-huh.

C: One more question. What does the "body" of wine mean?

M: Light-bodied is like skim, mid-body is like whole, and full-body is like cream.

C: Got it.

M: I knew you would.

40. WHAT IS THE BEST WINE TO GIVE AS A GIFT OR BRING TO A PARTY?

There are a few questions you should try to answer before going into the wine store. First and foremost is how much do you plan on spending? There are quality wines for every budget, and believe me, if they are truly wine lovers, they will be happy with any wine you give them.

Next question: Does the gift recipient "know" wine? If they do, a Cabernet is a safe bet. Unless you have a favorite Cab, I would look for one with a high rating. If the bottle got a high score, it would probably say it on the shelf talker, which are the small, printed tickets attached to retail shelves. If you're choosing wine based on a shelf talker, make sure the rating is for that vintage of that bottle. There are also numerous websites that show wine scores. Winesearcher. com seems to have emerged as the industry standard. Keep your phone handy and look 'em up!

If your gift is for that hoity-toity person in your life, you may want to go for a French or Italian wine. For a French wine, pick any Bordeaux that fits your budget. If it says *Chateau*-something, it's probably Bordeaux. If it says *Domaine*-something, it's probably from Burgundy and is likely more expensive.

For an Italian wine, look for a Ripasso. That's a process, not a wine—a laborious and expensive process that concentrates the grapes. At the high end is Amarone. That's the ultimate Ripasso. But there are winemakers in Verona's Valpolicella region that use the leftover skins of the process to make a Valpolicella Classico or "baby Amarone" at a third of the price: a definite crowd-pleaser. If money isn't an issue, *ah salute!* Then go for a bottle of Barolo or Brunello. Hopefully, they will return the favor when it's their turn to give you a gift!

If the recipient is not necessarily a connoisseur but enjoys wine every now and then, I'd stick to well-known brands—Kendall-Jackson, J. Lohr, Joel Gott, Josh, Caymus. These wines are available everywhere and please a wide range of people and palates, which is why they *are* so well known.

On the wine floor, I like to recommend blends. It's a perfect choice when you have no idea which wine grape they may prefer. (To read more about blends, go to question #44.)

An alternate route is to find out their favorite food. Knowing the person loves BBQ, sushi, or steak is a clue to picking the perfect wine. I love helping customers pair wine with food.

Lastly, any kind of bubbly makes a nice gift, if only because people rarely buy sparkling wine for themselves. Champagne is impressive if it's in your budget, but a Prosecco or Cava is a perennial crowd-pleaser and is more affordable.

There's always a chance the lucky recipient will be so appreciative of your choice that he or she opens it up and shares a glass with you!

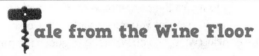

Tale from the Wine Floor

Customer: Can you help me pick wine presents?

Me: Sure.

C: Here's my list. The number next to each name is how much I wanna spend for that person.

M: I'm curious, how did you decide who is worth more and who is worth less?

C: Simple. If I like them, I spend more. If I don't, I spend less. But I'm looking for wines on sale, so they think I spent more.

M: So that they think you like them, even though you don't.

C: Right.

41. WHAT AND HOW MUCH WINE WILL I NEED FOR A PARTY OR WEDDING?

When you're deciding on wine to be served at a large party or wedding, there are a few things to consider. First, how much do you need? A 750 mL bottle of wine has 25 ounces—that's five glasses. A 1.5 L is double that. Most stores will allow you to return unopened bottles, so take into account that if you choose the larger bottles, you may not be saving money.

The rule of thumb is one glass per person per hour. Be careful when you are trying to figure the percentage of wine drinkers. The "my friends drink beer" logic can get you into trouble. That being said, not everyone will be drinking wine, so you will need to guess the percentage of guests who will—the average is 50%. The rest are the heathens that drink liquor, beer, or seltzer. (That's my favorite sentence in the book.)

Let's use a four-hour reception example with 150 guests. 75 guests x 4 glasses = 300 glasses. Divide that by five (the number of glasses per bottle) and you have 60 bottles (750 mL). You can forgo the math altogether by rounding up and figure one bottle per person. So, you have a few "wedding bottles" to drink on each anniversary for a couple of years. Not a problem.

Next, you'll need to decide what kind of wine you would like to serve. Unless you are in the wine industry, your guests are not expecting a wine-tasting event. Pick one white and one red. If you were to have two or three varietals of whites and reds, you run the risk of running out of at least one of the wines. (It's not the end of the world, but running out of anything at a wedding is a no-no.) If you prefer a slightly sweet white, make the other a dry red. Or the reverse: If you choose a sweet red, the other should be a dry white.

For white wines: I do not recommend a wine that's too sweet, like Moscato, because it's such a terrible match with any food except dessert. Riesling is a good choice. It has enough sweetness to please a sweet-wine drinker yet pairs well with food. A good dry white is a Pinot Grigio (the mainstream choice) or Chardonnay.

For reds: There are a lot of semi-sweet red blends available. Sweetness is subjective. Try a couple before deciding. A good dry red choice is Merlot (the mainstream choice), Pinot Noir, or Malbec. All three will please most red wine drinkers. Cabernet Sauvignon can be a good choice, but it has too many variables in style. Merlot has a similar flavor profile as a Cab, but it has fewer tannins, making it "softer" and more appealing to a wider age range of guests.

If you're preparing for a wedding, remember a toast may be required. You will be able to get about eight glasses from a sparkling bottle that is poured in flutes. If you can serve real (meaning French) Champagne, good for you. It is rare to have the real stuff poured at most events due to its price. Fortunately, a Cava from Spain or Prosecco from Italy will do just fine and is a terrific value. Keep in mind that you will need a glass for every guest, and some might continue drinking it (like me!), so add a few bottles after your calculation.

Important note: Instruct your venue not to pour the sparkling wine too early and have it sitting on the tables getting warm during the cocktail hour.

Lastly, keep in mind that it's *your* party. If you love a certain wine, forget all the rules and serve it on your special day.

Tale from the Wine Floor

Customer: I have an unusual request.

Me: It's OK, I get a lot of them. What can I help you with?

C: I'm getting married, and I need to pick out wines for the wedding, but I don't want them to be too good.

M: I understand, you don't want them to be too expensive.

C: No, I don't want them to be too good!

M: Why not?

C: Well, if they're too good, people will drink a lot. I want the wine to be just good enough that they drink it, but not so good that they'd drink too much.

M: Got it. OK, this wine sucks, but it's not as bad as this wine. And this wine is the worst of these three bad ones.

C: Is it a bad wine, or you just don't like it?

M: I don't like any of them.

C: But they sell, right? So, somebody must like 'em!

M: Right. Sounds like we have a winner.

C: Now, where is your beer section?

M: Good beer or bad beer?

C: You're a funny guy. I should invite you to the wedding.

M: I'd come, but I heard you're serving bad wine.

Tale from the Wine Floor

Customer (with foreign accent): You help, yes?

Me: I'd be happy to. Yes.

C: Wine for wedding. This wine. You like?

M: Well, it is popular. But I don't think it's a good choice because of the name.

C: What means Ménage à Trois?

M: It means "not appropriate for wedding"!

42. WHAT IS THE BEST WINE TO COOK WITH?

Let's start with a few ground rules: the first is not to use wine called "cooking wine"! If you don't want to drink it, you shouldn't cook with it. Secondly, you can add sweetness but can't take it out. So, unless the recipe calls for sweet or fortified wine, stick with dry wine varietals. In the red wine category, good choices are Merlot and Pinot Noir. Stay away from highly tannic wines that can leave unpleasant bitterness in the dish. For whites, Pinot Grigio, Sauvignon Blanc, and Chardonnay (un-oaked) are good choices. And don't forget sparkling wine! Although the popular choice is to create a Champagne vinaigrette or sorbet, it also makes a great Beurre Blanc sauce. The bubbles dissipate during cooking, and you can even use the flat bottle you didn't finish last night. (Yeah, like *that's* ever happened to me.)

After you've chosen the wine you'll use while you cook, you'll need to consider when you should add the wine. You can use wine early in the cooking process if you need a large quantity that will cook down, say for beef bourguignon or risotto. Deglazing comes at the end of the process when you need a small amount added to the pan of stuck-on bits called fond. It adds richness, acidity, and depth of flavor.

Useful tip: You can freeze wine in ice cube trays, making it the perfect solution when you just don't have the right wine handy for cooking. Be sure to cover the tray with plastic wrap to stop ice crystals from forming. Another tip is to use the larger silicone ice cube trays (made for drinking Scotch) because each cube usually equals one-half cup.

Fortified wine has many cooking applications. Here are a few:

- *Marsala*: Think outside the pan and use this classic fortified wine in dishes other than chicken or veal Marsala. It works in all kinds of braised meat preparations.
- *Sherry*: Just a dash added to soups and cream sauces brings a layered dimension to the dish. It is also great for deglazing.

- *Madeira*: There are several styles of this distinctive wine. Choose the dryer, lighter Sercial. It can also be used as a substitute for Sherry in any recipe.
- *Port*: Try just a splash of Ruby style in your favorite spaghetti sauce. Or just reduce a little with a pat of butter and serve over steak, lamb, or duck.

43. WHAT ARE BOX WINES, AND ARE THEY ANY GOOD?

Wine in a box is actually wine in a bag. The polyethylene bladder inside the box collapses on itself as the wine is dispensed, keeping oxidation at bay. It was the invention of South Australian winemaker Thomas Angove. He was given a patent for his "bag in a box" on April 20, 1965. It was improved upon in 1967 by another Australian, Charles Malpas, and Penfolds Wines, by adding an air-tight tap, which made it even more convenient by not requiring constant resealing.

Are box wines any good? They are good in that they serve a purpose. (I know, I went for the safe answer on this one.) Actually, box wines are the perfect container for a college dorm room or beach, pool, or backyard party. They are portable, unbreakable, easy to open, and they last longer than bottles after opening. What more could you want? Well, let's take a look at some pros and cons of wine in a box.

Pros

- They are portable. There's even a built-in handle.
- They're easy to open. No corkscrew is needed.
- They have a shelf life of 30 or 40 days compared to 2 or 3 days for most bottled wines since the bag system offers less air-to-wine contact.
- They are environmentally friendly. Cardboard and sustainable bags pose less of an environmental impact than glass bottles.
- They are convenient, so you can have a glass every now and then when the spirit moves you.
- They're handy to use in cooking since you don't need to open (or keep open) entire bottles when all you need is a half cup for a recipe. For that reason alone, nearly all restaurant chefs use box wines in their cooking.
- They are a good bang-for-your-buck. Most boxes are equivalent to four bottles of wine.

Cons

- The quality of the wines is made for, let's say, a less discerning wine palate. Most boxes are sweet wines, unusual blends, or basic versions of the most popular varietals.
- There is a limited selection. If you are not a fan of sweeter wines or most popular varietals, you're out of luck in most markets.
- There is no age ability, which means there is no chance for a box wine to get better. Most boxes even have an expiration date!

Aside from the aging dilemma, I believe there is no reason why a wine box can't have better-quality wines put inside them. Brands like Black Box and Bota Box have introduced premium wines in a box with the hope of beating the stigma that boxed wines are only of lower quality. Statistics show that the average drinkers of box wines are newbie wine drinkers and senior citizens. That could be attributed to the value of box wines or to the wine palettes of these two groups. At the time of this writing, sales of box wine have surged over 53%, although that is believed to be due, in large part, to the pandemic of 2020.

So, if you want my opinion as to whether wines from a box are good, my answer is not really *right now*. Will they be better in the future? It's possible.

Tale from the Wine Floor

Me: Can I help you?

Customer: Nope. I know everything there is about wine!

M: Good for you. I'll be right here if you . . .

C: I'm an afficiato! In fact, I'm what you call a *somminair*.

M: A what?

C: A SOMM-IN-AIR. That's what they call a wine expert.

M: Oh. Got it. I never met a "SOMMINAIR." Well let me know if I can tell you where things are.

C: OK, where are your box wines?

44. HOW DO I KNOW IF IT'S A BLEND?

This is a tricky answer because there are many caveats and inconsistencies for what should be a simple definition. A blend is a wine that uses more than one grape. How can that be misunderstood? I mean, it states it right there on the label, right? Not so fast! Sometimes a wine that you think uses a single grape is also a blend! That Cabernet you are holding only has to have 75% of the grape in order to state on the label that it's a Cab. In the European Union the requirement is 85%.

A wine that uses a few different grapes isn't something new. Bordeaux, Port, Champagne, Rioja, Chianti, and Châteauneuf-du-Pape are all classic blends, but there are other, let's say, unique blends that are anything but classic. I call them "kitchen sink" blends. These modern blends will have a name that isn't a grape or a place—just a catchy name. "The Prisoner," "Apothic Red," "Conundrum," "Tapestry," etc.—all blends. Some just say "red wine"! I guess they couldn't think of a catchy name.

The grapes are sometimes listed on the back label but are often proprietary, meaning the winemaker doesn't want you to know, so they have the option of changing the blend year to year. That's not necessarily a bad thing. Confusing, but not bad.

Blending, in the hands of a good winemaker, can add complexity and even enhance the elements of the main grape. It also allows the winemaker to make a totally unique wine, combining aromas and textures that a single grape may not be able to offer from that region. Some grapes are understood to be in the supporting role (Example: Petite Verdot in Bordeaux). After all, harmony requires more than just the guy with a low voice to sing the bass part. You need someone to sing soprano, tenor, and baritone. In this song analogy, think of the dominant grape as the melody.

Want to try your hand at winemaking? Next time you have a little left in this bottle and a little left in that, blend them together. I guarantee you'll have a new respect for winemakers and may even invent a new blend: Chateau You!

$\mathbf{\text{T}}$ale from the Wine Floor

Customer: I'm looking for this wine I had and liked.

Me: What was the wine?

C: It was a blend of Cabernet and Sauvignon.

M: OK, well, ya see, Cabernet's whole name is Cabernet Sauvignon. It's not a blend. That's one grape. If you'd like to get geeky about it, it was originally a cross of Cab Franc and Sauvignon Blanc.

C: So you don't have it?

M: We have a whole aisle of them. I was just explaining that it's one wine, not a blend.

C: So what's Merlot's second name?

M: Merlot doesn't have a second name.

C: So they don't blend that?

M: They do! It's in the classic blend of Bordeaux.

C: What's it blended with?

M: Cabernet.

C: And Sauvignon?

45. HOW MUCH SHOULD I SPEND FOR A GOOD BOTTLE OF WINE?

This is surprisingly one of the most difficult questions that a wine professional gets because answering it forces you to assume so much. Wine is, after all, as subjective as any other food or beverage. Not having the faintest idea if you like chocolate or vanilla, drink Coke or Pepsi, or take your coffee with or without cream, the best way to answer this is to explain when wine takes a step forward in quality and what that step costs.

- **Extreme Value** ($5 to $10): These are the bulk wines made by large commercial wineries that use grapes from large regions. Here, price rather than quality is the main objective. Whether it can *sell* is the only concern.
- **Value** ($10 to $15): Although price is still the main focus, value wine can come from under-valued regions and unknown grapes.
- **Premium** ($15 to $25): This is the first step in wine "typicity," meaning the wine is varietally correct and tastes typical of the grape it's made from. Not necessarily a high bar, but it is the beginning of QPR (quality-to-price ratio).
- **Super Premium** ($25 to $50): While still keeping an eye on the market, the wines in this category can be cellar-worthy, highly rated, and in demand.
- **Luxury** ($50 to $100): There's no way that a wine lover can't find an excellent wine at this price point. There's also no way that the average person will spend it, which is OK since there's not a lot of it to go around.
- **Super Luxury** ($100 and up): This is when wine gets silly. Are these wines the best of the best? Probably. Would I buy them? I have on rare occasions, but no.

Tale from the Wine Floor

Customer: I have a picture on my phone of a wine I bought here.

Me: (looking at phone) I'm sorry, but we don't carry that wine.

C: But I'm sure I bought it here!

M: No disrespect sir, but you're mistaken. We don't sell that wine.

C: Can you check the computer or something? Maybe you're mistaken!

M: Fair enough . . . I'll do that. Maybe we had it before I started here.

(5 minutes later . . .)

M: OK, I'm sorry sir, I checked the computer and went to the accounts department and talked to the person who enters all products into the system and we never in the 60-plus years this store has been here ever sold that wine. I would be happy to recommend a wine with a similar grape and region if you'd like.

C: No, I don't want another one. The tag says $7.99, and I was charged $15.99. Who do I talk to about the difference you owe me?

Tale from the Wine Floor

Customer: Can you help me? I'm looking for a bottle of wine for a friend that will show the price as 50 bucks if they Google it.

Me: This one.

C: Great, how much is it?

M: 50 bucks.

46. WHY DID AMERICA START AND STOP PROHIBITION?

The 18th Amendment to the Constitution, which prohibited the sale of alcohol, became law at 12:01 a.m. on the 17th of January 1920. Some states refused to enforce the law, and other states, like Maryland, never enacted it in the first place—but not for the reasons you may think.

Alcoholism undoubtedly was a problem. The average American was drinking 27 bottles of hard liquor per year; that's 14 times more than today. Amazingly, anyone over 15 could drink. However, there were other social problems that Prohibition intended to fix. Child labor laws, poverty, and even anti-immigrant views all had a part in what was called the "Noble Experiment."

Following World War I, the vast majority of the public was uneducated and thought the National Prohibition Act, informally known as the Volstead Act (named after the sponsor, Andrew Volstead), would still allow lower-alcohol drinks like beer and wine. Upon the realization that the new law would forbid the manufacture, sale, and transportation of liquor—not their consumption—a buying frenzy began that guaranteed no one who wanted a drink did without. A staggering 141 million bottles of wine were sold during the three months prior to the starting date.

In addition, it was (and still is) legal to produce up to 200 gallons of wine per year for personal use, which comes out to a little more than two and a half bottles per night, per household. To make the process easier, enterprising winemakers made and sold a wine "brick" of concentrated grape juice, which came with the following "warning": "After dissolving the brick in a gallon of water, do not place the liquid in a jug in the cupboard for twenty days, because then it would turn into wine." Ironically, US consumption of wine rose during Prohibition from 70 million gallons per year to 150 million gallons per year.

Although most proponents were initially optimistic about the prospects of the legislation, they soon realized that Prohibition allowed businesses to exploit loopholes in the law. Drugstores were allowed to sell "medicinal" whiskey to treat even the most common ailments, like a toothache. Walgreens, a drugstore founded in 1901, took full advantage of this and grew from 20 locations to over 500 stores. There were over 30,000 illicit establishments known as "speakeasies" (so named because of the practice of speaking quietly so as not to alert the police) in New York City alone.

A few wineries, including Louis M. Martini, Beaulieu Vineyards, Concannon Winery, and Beringer Winery, stayed open, citing sacramental wine for Catholic Mass as their sole purpose of production.

Prohibition came to an end on December 5, 1933, when the 21st Amendment was ratified. The 18th Amendment was the fastest US amendment to be created and repealed. There were as many reasons for repeal as there were for ratification. The nationwide shock that followed Chicago's Saint Valentine's Day Massacre in 1929 is an oft-overlooked contributing factor to the amendment's repeal, as Prohibition was supposed to reduce crime. But the most powerful argument for repeal was the stock market crash of 1929 and the Great Depression that followed. Legalizing booze was an obvious opportunity for much-needed tax revenue. Curiously, some regions of the country maintained a ban on alcohol even after repeal. Oklahoma remained dry until 1959, and Mississippi was alcohol-free until 1966.

In the years following the repeal of the Volstead Act, American wineries were forced to ramp up production beyond their capabilities. They were also trying to supply types of wines that were familiar to the soldiers returning from war—like Burgundy, Chablis, and Champagne—which the United States could not produce. American barrels at the time were made of redwood rather than oak, which had atrophied during 13 years of non-use. They leaked and oxidized the juice and produced terrible wine. So, when entrepreneurs like Ernest & Julio Gallo started producing wines of better quality, Americans drank away, and jug wines became the norm. But the main and lasting effect of Prohibition is what is called the "Three Tier System." During Prohibition, distribution of alcohol was handled by the criminal underworld. Elected officials understood that they would be putting the fox in charge of the henhouse if a system wasn't put in place to ensure market conditions and, more importantly, control taxes. The resulting system mandated that alcohol manufacturers could not sell directly to the public. They must sell to a middleman, the distributor.

Here's how it works to this day:

1st Tier—The Supplier/Producer
- Makes the product.
- Pays federal excise tax.

2nd Tier—The Distributor/Wholesaler
- Purchases the product from the supplier, then sells it to retailers/licensees.
- Pays state excise tax.

3rd Tier—The Retailer/Licensee
- Sells alcohol "on premise" (restaurants) and "off premise" (liquor stores).
- Pays state sales tax.

The Takeaway

The consequences of Prohibition are hard to quantify. In monetary terms, Prohibition cost the federal government an estimated $11 billion in lost tax revenue while costing over $300 million to enforce. In addition, it all but created organized crime in America and made folk heroes of gangsters like Al Capone, Lucky Luciano, and Bugsy Siegel. On a social level, it gave rise to the *us-against-them* mindset and mistrust of the federal government that is evident even after 100+ years since the "Noble Experiment."

ale from the Wine Floor

Customer (sitting on box): This store is huge. You should make it easier for people to find what they're looking for.

Me: We try to, but I'm happy to help you. What are you looking for?

C: Lambrusco.

M: You're sitting on it.

Chapter Four

PAIRING & SERVING

 ale from the Wine Floor

Customer: Does this wine taste like a cupcake?

Me: Uh, no. That's just the name of the producer. Cupcake makes all kinds of wine.

C: But none that tastes like a cupcake?

M: No. But there are wines that pair well with cupcakes.

C: That means when you're eating cupcakes?

M: Yes, exactly.

C: So shouldn't somebody just make a wine that tastes like cupcakes?

M: I guess that makes sense.

C: OK, so which wine should I drink if I were eating a cupcake?

M: This. It's a Moscato.

C: Does Cupcake make one?

M: Yes!

47. IS THERE A SCIENCE TO WINE-AND-FOOD PAIRINGS?

"The apple of my eye," "a bitter pill to swallow," "sour grapes," and "taken with a grain of salt" are all clichés that demonstrate the connection taste has with our emotions. But it also has to do with evolution. A bitter taste was once an indication that the food was poisonous, whereas the taste of sweetness was a sign of nutrients and pleasure. Yet, the science of how certain flavors are perceived is still debated.

It was universally accepted for years that there were four basic tastes at the root of our enjoyment of food: sweet, sour, salty, and bitter. However, in 1910 a Japanese scientist named Kikunae Ikeda described the savory taste of glutamates, which are characteristically found in broths and cooked meats, that he believed didn't fit into any of the four categories. It wasn't until 1985 that the term "umami" was recognized as the fifth taste perceptible by the human tongue.

The century-old idea that different areas of the tongue are associated with each taste has also now been debunked. Scientists now believe that the entire tongue can taste all of these flavors. Taste buds, those small bumps on the tongue called *fungiform papillae*, are made up of 50 to 100 receptor cells. These cells sense and relay the sum of sweet, sour, salty, bitter, and umami to the brain. As discussed in #26, "How can I expand my wine palate?," perceptions of taste are influenced by the number of taste buds a person possesses, their culture, and their upbringing. So, foods don't taste the same from person to person.

What do our taste buds sense in a glass of wine? A single glass of wine contains thousands of chemical compounds. These compounds are not only determined by the grape variety itself but by the soil and climate in which the grape is grown. Wines have acids (tartaric, malic, citric, and lactic) that activate the salivary glands and help enhance the flavors of food. Wine is also typically served in a specifically designed glass, which concentrates the aroma and activates olfactory receptors in the nose. It is the interaction between the taste buds and olfactory receptors with the acids and other volatile components that comprise the sensation of taste.

While this explanation would seem to make the pairing of wine with food a scientific certainty, any wine expert worth their salt will tell you the best wine-and-food pairing is the one you like! Pairing preferences are subjective. If you like a Cabernet with tuna, have at it. But there are pairing rules that are universally accepted and worth a try.

General Rules

- The weight of the wine should match the weight of the food.
- Protein will soften tannins.
- The acidity of the wine should match or exceed the acidity of the food.
- Fish oils love acidity but hate tannins.
- Acidity cuts saltiness.
- Alcohol enhances spice, which can add to the heat.
- Sweet tames heat.
- A dessert wine should be as sweet or sweeter than the dessert.
- When in doubt, match the color of the food with the color of the wine.

Menu Pairings

Appetizers
- Chips & Dips: Sparkling wine or Rosé
- Soups & Salads: Sauvignon Blanc, Albariño, or Verdicchio
- Charcuterie: Dry Lambrusco, Prosecco, or Fino Sherry

Main Courses
- Turkey: Dry Riesling, Beaujolais, or Pinot Noir
- Beef: Cabernet or Syrah/Shiraz
- Meat stews: Côtes du Rhône or Bordeaux
- Chicken: Sauvignon Blanc or Beaujolais
- Pork: Riesling or Merlot
- Lamb: Pinot Noir or Rioja
- Seafood: Chardonnay or Chenin Blanc
- Fish: Pinot Grigio or Albariño
- Sushi: Prosecco, Gewürztraminer, or Riesling
- BBQ: Zinfandel or Syrah/Shiraz
- Pasta with red: Chianti, Barbera, or Valpolicella
- Pasta with white: Pinot Grigio or Chardonnay

- Vegan dishes: Sauvignon Blanc, Pinot Gris, or Grüner Veltliner
- Quiche/Omelets: Beaujolais Villages, Sauvignon Blanc, or Cava

Desserts
- Chocolate: Ruby Port or Banyuls
- Fruit: Moscato or late-harvest Riesling
- Nuts: Sherry, Tawny Port, or Vin Santo

ale from the Wine Floor

Customer: Does this wine go with steak?

Me: Not really. That's Moscato. It's sweet. It would be like putting sugar on your steak.

C: But I only like sweet wine.

M: Then it doesn't matter what it pairs with. Drink what you like.

C: So, what does go with steak?

M: A red wine normally.

C: So, red Moscato?

M: Perfect. A match made in heaven.

48. WHAT WINE PAIRS BEST WITH THANKSGIVING DINNER?

Ahh, Thanksgiving dinner: the ultimate culinary labor of love. It's the pull-out-all-the-stops meal that home cooks plan and prepare weeks in advance. Unfortunately, when it comes to wine pairings, there is often little advance thought, much less planning. Most people stick with the wine they always drink, even if it doesn't pair well with turkey. But have no fear! You can be a little more adventurous and a bit more festive without interfering with family traditions or tight holiday budgets.

I recommend starting with a bubbly. (OK, I always recommend starting with bubbly.) It sets the festive mood and matches every course, from chips and dips to entree and dessert. An extra dry or demi-sec is the way to go. They have a touch of sweetness—which you're going to need as there is a lot of sweetness on a traditional Thanksgiving plate. This might be the perfect thing for a favorite aunt or uncle to bring. They might even think they're getting off easy, and they are! There are sparkling wines at every price point, from bang-for-the-buck Cavas from Spain and trendy Proseccos from Italy to fancy-schmancy Champagnes from France. You can find a bubbly version of nearly every grape nowadays.

Now, let's say you're having a bigger party. More attendees means more people to help out with the wine. In this case, it's probably a good idea to give a little direction to the guest who wants to help. "Aunt Reen is bringing Prosecco. Would you like to bring another kind of wine, maybe a Riesling?" Crafty, right?

Now, if you want to go all out, put together a group of wines that not only matches your carefully and lovingly prepared meal but also pleases the widest range of palates, from your keto cousins to your gluten-free friends. If you're going this route, keep the bubbles and Riesling, but add a Beaujolais or Rhône blend. Beaujolais would be my choice. It's different enough without being weird, fruity enough without being a fruit bomb, readily available, and reasonably priced. I suggest a Beaujolais-Villages or, if your budget allows, a Cru Beaujolais. (See #77, "What's the difference between Beaujolais and

Beaujolais Nouveau?") The other choice, a Côtes du Rhône, blends a few grapes, mainly Syrah and Grenache. Don't even worry about the producer; the entire region produces wines that are flat-out bargains.

And finally, if you are like my family and have an entire dessert table, you should consider a dessert wine. Port is the perfect pairing with every dessert on the table. Ruby is the classic wine for all things chocolate, and Tawny is the quintessential wine for fruit and nut pies or a cheese platter. Once opened, Port can last throughout the holiday season.

There you have it. Wines that please the masses yet pair well with the smorgasbord that is the traditional Thanksgiving plate.

Tale from the Wine Floor

Me: Can I ask you something?

Customer: Sure.

M: I'm asking everybody. What will you be drinking for Thanksgiving?

C: Fireball and RumChata.

M: I mean with your dinner.

C: Fireball and RumChata.

M: Sounds perfect.

49. WHAT IS MULLED WINE?

A mulled wine is a beverage made with red wine, mulling spices, and raisins. The combination of "mulling" spices varies but usually consists of cloves, nutmeg, star anise, cinnamon, pepper, and cardamom. It is traditionally served during the Christmas holiday season. The drink is especially popular in the United Kingdom, but nearly every country in the world has a version of mulled wine—the most popular being Glühwein from Germany.

It's unclear exactly when and how the first mulled wine came about and who made it, but the first recipe appeared in *The Forme of Cury*, a medieval cookbook from England, in 1390.

If you'd like to try it, here is a traditional recipe. Put a pot on the stove, and add Christmas music (brass quintet preferably) and candles. It pairs best with people you love.

Traditional Mulled Wine Recipe

These are only ingredients and guidelines. Make it your own. Just remember to write down what you added so that you are able to make it again!

Ingredients
- 2 bottles of red wine. Use a full-bodied wine you like. I recommend Zinfandel, Grenache, Merlot, or Barbera.
- 1 orange studded with 6 cloves
- Zest from 1 lemon
- 2 cinnamon sticks
- A couple of star anise
- A touch of honey
- A touch of grated nutmeg
- ½ cup brown sugar
- ¼ cup triple sec or other orange liqueur
- A few orange slices

Instructions
1. Pour the wine into a large saucepan.
2. Add the orange and lemon zest, spices, and sugar, and heat gently until simmering.
3. Reduce the heat as low as possible, and simmer for about 30 minutes, so the spices infuse.
4. Add the orange liqueur and orange slices.
5. Ladle into small cups or glasses to serve.
6. If you'd like to kick it up, add a touch of brandy or vodka.

50. WHAT DOES DECANTING ACTUALLY DO?

Decanting is a fancy word for pouring the contents of a bottle of wine into another container. There are two reasons for the practice: First, simply opening a bottle to "let it breathe" does very little (if anything) since the surface exposed to air at the top of a bottle is about the size of a penny. Therefore, decanters have a shape that widens that surface. Think of cutting a slice of an apple—in a very short amount of time, the exposed flesh will start to brown, yet the rest of the apple won't. Decanting exposes all of the "flesh" to the air. The process of pouring the wine into a decanter itself will tone down some acidity, blow off any "off" aromas from the bottle, and soften the astringent tannins of a young wine.

It's one of wine's biggest ironies that oxygen is the enemy of wine throughout the entire winemaking process, and then we decant and swirl the heck out of it to give it air before we drink it. There are even people who use a kitchen blender! No, I haven't tried it. I think it's too extreme . . . and disrespectful.

The second reason for decanting is to separate the sediment from the rest of the wine so that the gunky particles of dead yeast cells, tartrates, and polymers stay at the bottom of the bottle—not that they are harmful, but they taste bitter and make the wine cloudy. Modern filtering techniques have made sediment less of a problem than it was years ago, which makes decanting necessary only for aged Ports and wines with substantial tannins. However, there are wine lovers that like all of their wines decanted, even Champagne!

There are certainly some wines (Champagne included) that should not be decanted. Aerating a fresh and fruity Rosé or mass-marketed "value" wine is possibly the height of pretense. Old red Burgandy, on the other hand, could actually dissipate and taste worse after decanting! If you're confused, welcome to the club. I would suggest you try the wine after opening. If the taste is muted or just isn't what you were expecting, it couldn't hurt to decant it. And if you'd rather pour from a bottle than an awkwardly shaped decanter, you could pour the wine back into a rinsed-out bottle. Ta-da! You have double-decanted. Aren't you fancy?

51. DOES SERVING TEMPERATURE REALLY MATTER?

From ice cream to pizza to a simple cup of coffee, serving temperature matters. Studies show that overall perception of taste increases and decreases at specific temperatures on a molecular level. It's not surprising, then, that wine—with all its complexities—changes dramatically if served too hot or too cold. Serving it too cold masks the fruit and elevates the acidity. Serving it too warm accentuates the tannins, and alcohol levels will completely overwhelm your palate . . . and your nose!

Many people may have heard that red wine should be served at "room temperature." That made sense a hundred years ago, when rooms were a lot colder. Today most homes are heated to about 74°F or more. (Growing up, touching the thermostat dial was forbidden in our house. It was set at 68° and stayed at 68°. I also distinctly remember my mom yelling for us to close the front door. "What are we trying to do, heat the whole neighborhood?" But I digress. . . .)

Knowing that most wine drinkers will not have a dedicated wine fridge, the two obvious ways of getting a bottle to the correct serving temperature are a refrigerator or an ice bucket.

I recommend a 20-minute rule: Leave your white wines in the refrigerator and take them out 20 minutes before drinking. Leave your red wine out but put them in the refrigerator for 20 minutes before drinking.

If it's an exact temperature you're looking for, here's the "Goldilocks" zone:

Serve red wine between 62° and 68° Fahrenheit (~15° to 20° Celsius) and serve white wine between 49° and 55° F (~7° to 12° C).

52. WHAT IS THE BEST WAY TO CHILL A BOTTLE OF SPARKLING WINE?

The bottle should be chilled to a temperature between 40° and 45° Fahrenheit. Leaving the bottle in the refrigerator for a couple of hours will get it there. You can also put a bottle in a freezer for 15 minutes. Just remember to set a timer. There's a good chance you'll forget that it's in there, and every chance that the bottle will explode.

The classic way to chill a bottle is to place it in an ice bucket filled with half ice and half water. Leaving it for 15 minutes should do the trick. Adding a cup of salt to the bucket will speed up the process. Sodium chloride lowers the freezing temperature, allowing the water to become colder than 32° Fahrenheit without turning to ice.

Most people believe that a bottle that overflows immediately after opening is the result of someone inadvertently shaking it. That may be the case, but a more likely reason is that the bottle wasn't cold enough.

Tale from the Wine Floor

Customer: I'm looking for an Asti Champagne by Cupcake.

Me: Well, an Asti is from Italy and Champagne is from France. Cupcake does make a Moscato D'Asti, which I think you mean because it's Frizzanti, but it isn't an Asti or a Champagne.

C: If it's fizzy, it's like a Champagne though, right?

M: Only if you think a 7Up is also like a Champagne.

C: I like 7Up, so which Champagne tastes like that?

M: No Champagne does, but I guess Moscato d'Asti by Cupcake is close.

C: Isn't that what I asked for?

53. WHAT IS THE CORRECT WAY TO OPEN A BOTTLE OF SPARKLING WINE?

Take off the foil cap and immediately put your thumb on the top, and don't take it off! Unscrew the cage (it's always six turns). Wrap your hand around the top, cage and all, and turn the bottle—*not* the cork. Push back when you feel the pressure. There should be a slight hiss, not a pop.

Fun fact: A raisin that is dropped in a glass of sparkling wine will continuously go to the bottom and back up to the top of the glass. The reason? As bubbles form in the crevices of the raisin, the buoyant force brings the fruit to the surface. Once the bubbles dissipate, the raisin will sink to the bottom, and the process repeats.

54. DOES USING THE CORRECT WINEGLASS REALLY MATTER?

It's easy to state the obvious that what is in the glass is all that matters. But there is a tipping point between what is unpretentious and what is ridiculous. Sure, most people don't routinely use the expensive glassware they were given as a wedding present, but do you really want to drink wine—this liquid work of art—out of a white Styrofoam or red Solo cup? So let's look at what matters and what doesn't.

Cab, Bordeaux, Zin, and Syrah are typically high in alcohol and tannin. Having a glass with a larger bowl will help dissipate ethanol and allow more oxygen to soften the tannins. On the other hand, aromatic whites do not have tannins and benefit from glasses with a narrower bowl to concentrate the aroma and reveal fresher fruit—not to mention, it keeps the juice cold longer than those with large bowls.

Sparkling-wine glasses are more about aesthetics. Seeing the continuous streams of bubbles in a flute is pretty cool, and (they claim) it keeps the effervescent flow longer. (I prefer drinking Champagne out of a traditional white-wine glass.) Fortified wines, with their higher than usual alcohol content and concentrated flavors, logically require a smaller glass because the serving size is smaller. Plus, a glass with a large bowl would completely overpower your nose.

Stemless glasses *do* have a purpose. Everyday wines don't need all the introspection and hoo-ha, just pour a glass and pass the pizza. Having a stem just makes it easier to knock over a glass. Just know that the warmth of your hand around the bowl will heat the wine—so drink faster!

Now, put on your nerd glasses for a few quick notes about the science of glassware: The shape of the glass affects how the wine hits your palate. It forces the liquid to travel to the back of your tongue or to the sides of your mouth. The larger the bowl of the glass, the greater the number of volatile and phenolic compounds—which are responsible for the aroma, taste, and mouthfeel of wine. The size of the rim precipitates contact with the lips at different points to make a more seamless transition of the wine to the tongue, thereby allowing it to focus on the wine and not the feel of the glass.

Still with me?

In short, the theory is that the shape of the glass affects not only the aromatics of the wine but also targets a particular part of the tongue. While there is some science behind that theory, there is also hype. So, let's just say the same wine *can* taste different in a different glass.

Is there one universal glass that feels right, is not too heavy and not too thin, and serves most wine lovers' needs? Yes. But rather than advocate a specific brand, let me give you a few tips for purchasing your glassware:

- Pick a glass that has a stem that fits your hand without you having to use only two or three fingers.
- Be comfortable with the weight and balance of it. It shouldn't feel overly top-heavy.
- Make sure the size of the bowl has a wide enough base and has a slight narrowing at the top that allows you to swirl the contents without spilling easily.
- Google "lead-free, crystal, and dishwasher safe wineglass" and let your budget be your guide.

55. WHAT IS THE CORRECT WAY TO CLEAN A WINEGLASS?

It is perfectly safe to put wineglasses in the dishwasher. No, really! Just make sure they have enough room so there is no chance of them clinking together during the wash cycles. But more times than not, I wash them by hand. Keep in mind there is no need to put a lot of pressure or effort into it. It's not like washing the windshield of your car. Take it easy. Breathe.

Now, I have heard of adding everything from vinegar, hydrogen peroxide, baking soda, club soda, even Fixodent to the water when washing wineglasses, but here is what I found works best:

1. Hold the glass by the bowl, not the stem, and wash it in a sudsy bath using fragrance-free dish detergent. A sponge works better than a dish towel at getting lipstick off the rim.
2. Add a splash of bleach to a bowl of clean water, give the glass a dip, and then rinse thoroughly. It makes the glass crystal clear and disinfected. Don't worry, it doesn't leave a bleachy smell or taste.
3. Hand-dry with a lint-free, microfiber cloth or a basic flour sack—not your everyday cotton, terry-cloth kitchen towel.
4. The same method works on decanters, but you'll have to turn it upside down and air-dry it on a rack.

Note: I always give a quick smell to my glass before pouring wine into it. You would be amazed at how often a "clean glass" can have off-putting aromas.

56. HOW DO I HOST A WINE TASTING AT HOME?

I have been to literally hundreds of wine tastings. From large industry events with hundreds of people shuffling from table to table as though it were a frat party to small gatherings with buyers all whispering as if they were in a library. A little more casual but still in a work setting, salespeople will sample their latest and greatest product that they would like added to the shelf or wine list. I'm not complaining—it's not a bad gig. But my favorite way to taste and learn something new myself is by going to someone's home.

There are as many ways to do a wine tasting as there are wines. My recommendations are by no means the only way to do it, but they should make hosting your own party less daunting and seem more doable, which it is. Don't let your own expectations stop you from enjoying your friends in a whole new setting. That's really all that it is: a fun way to get together with friends over wine. Here are my thoughts.

The Theme

The easiest and least expensive way is to have a BYOB party where each guest brings a bottle of wine within a certain price range. That idea certainly works, but you may end up with too many bottles that are similar. You also leave yourself open for the inevitable moment when one guest dislikes a wine that another guest brought. Awkward! Picking your own theme is a better idea.

Here are a few themes to get you thinking on the right track.

- New-World Wines vs. Old-World Wines
- Wines of Italy: Tuscany vs. Piedmont
- Tiny Bubbles: Champagne vs. The World
- Pretty in Pink: A Rosé Bouquet
- Blends: Bordeaux vs. Meritage
- A Blind Challenge: Pick six favorite wines (three red and three white)

The Guests

Only invite friends who actually like wine! Having your cousin who says he drinks beer but would be happy to come along anyway changes the night. There is no need for any guest to be a connoisseur, but having at least a passing interest in wine isn't setting the bar too high.

Keep your guest count to a manageable number—10 to 12 people. It makes the party intimate enough for everyone to share in the same conversation and ensures that everyone is tasting the same wine at the same time. The size of your dining room table may dictate the perfect number. Having everyone spread out on different tables turns the night into a cocktail reception and makes getting everyone's attention a hassle.

The Setup

Professional tastings usually have a glass for each wine to be tasted. That is not needed at a home tasting. Two basic and inexpensive wineglasses, one for whites and one for reds, will work just fine. This is not the time to show off those delicate glasses you received as a wedding present.

You should have a "dump bucket" of some sort to pour out what's left in a glass after sufficiently tasting each wine—and plenty of bottles of water. Advise your guests to use both . . . unless you plan on combining your tasting with a sleepover.

Having papers and pencils keeps everyone on the same page, so to speak. Print the name and number of each bottle and leave enough blank space for tasting notes.

Scoring the wines is half the fun. A scoring key is helpful. Tally everyone's score at the end and declare a winner! You can use *Wine Spectator*'s 100 Point scale (see below) or invent your own.

Wine Spectator*'s 100-Point Scale*
> 95–100 Classic: A great wine.
> 90–94 Outstanding: A wine of superior character and style.
> 85–89 Very good: A wine with special qualities.
> 80–84 Good: A solid, well-made wine.
> 75–79 Mediocre: A drinkable wine that may have minor flaws.
> 50–74: Not recommended.

The Quaile Scale
    5 – Amazing. "Where can I get this wine?"
    4 – Great. "I really like this wine."
    3 – Good. "I'd buy it if I saw it."
    2 – OK. "I can drink it, but I wouldn't buy it."
    1 – Bad. "I don't like this wine at all."

The Quantity

You can get 12 two-ounce pours from one 750 mL bottle, so if you have 10 or fewer people, you probably will be OK with one bottle of each wine. Although having a second "just in case" bottle is a good idea. Six to eight different wines should make the party just long enough without having to Uber everyone home.

The Food

If you are planning on having this shindig at dinnertime, your guests may be expecting dinner! Having it a little later makes it perfectly logical to serve light hor d'oeuvres or cheese and crackers. If you would like it combined with dinner, the meal should follow the wine tasting. Personally, I don't mix the two.

What You Don't Need

Do not put flowers or candles on the table. The fragrance will throw off your sense of smell and make it difficult to distinguish the wine bouquet from the flower bouquet.

Final Thoughts

Remember to pour each wine in the correct order: Sparking → light-body whites → full-body reds.

I don't recommend having a sweet wine in the mix, but a dessert wine at the end is a nice touch. Just understand that you can't go back. After having something sweet, the dry wine you said you liked just minutes ago will taste awful. Then again, try it with the group. You'll see.

ale from the Wine Floor

Customer: How do I find a wine I like if I don't like wine?

Me: Let me see if I can help. What wines have you tried that you didn't like?

C: All of them.

57. WHEN IS A GOOD TIME TO HOST A WINE TASTING?

Picking a night other than a weekend keeps the party to a more exact time frame. I've seen a Saturday-night wine tasting turn into a beer and Bourbon tasting that lasted into the wee hours. All good if that's your plan; otherwise, two hours is about right. I'm thinking Thursday from 7 to 9 p.m.

On the other hand, if you're looking for a reason to celebrate, there are actually holidays devoted to wine. Every country has its own national wine holidays. Some are even called international (not that any other country was ever consulted).

Although Saint Vincent is the patron saint of wine, there are no sacred wine holidays. Funnily enough, he didn't make or even drink wine, but "Vincent" is a derivative of "vin-sang" or "blood of the vine."

America does seem to have as many holidays as wine grapes. Here are a few to celebrate if the spirit moves you.

- National Wine Day (May 25)
- Moscato Day (May 9)
- Prosecco Week (June 11–16)
- International Cabernet Sauvignon Day (August 30)
- Prohibition Repeal Day (December 5)
- And my favorite, obscure but true: Open That Bottle Night (Last Saturday in February)

Personally, I think it is only right to celebrate *all* of them. It's patriotic.

🍷ale from the Wine Floor

Customer: I'm looking for a wine that I only seem to find at the Capital Grille.

Me: Well, restaurant chains sometimes have their own wines.

C: Do you know where I can get it?

M: As a matter of fact, I do.

C: Where?

M: The Capital Grille.

Chapter Five

SAVING

58. HOW LONG DOES AN OPEN BOTTLE OF WINE LAST?

All wine, whatever the varietal or style, begins to change as soon as it gets exposed to air. That does not mean that it "goes bad" after a specific time. Some wines may actually get better. A little air will help the taste by softening tannins and "blowing off" some aromas to combine all the elements of the wine. That is, after all, why we decant a bottle and swirl wine in a glass. But catching an open bottle at the arc of its bell curve can be tricky . . . and subjective! I know some old-school Barolo lovers who won't take the first sip of the wine until it sits open for four days. For the rest of us, how long an open bottle lasts depends on the wine. A Cab or Zin, with high tannins and bold flavor, could last three or four days before you can detect a noticeable difference. Lighter styles, like a Pinot Noir, with lower tannins and softer fruit, may only last two or three days. Sparkling wines have a shorter time limit. They will start to lose their effervescence after the first day.

On the far end of the spectrum is fortified wine. Port will still be delicious for up to a month, while Sherry, Madeira, and Marsala can pretty much last indefinitely since they are already oxidized.

In all honesty, I'm not a fan of any preservation kits, argon gas canisters, or balloons that inflate inside the bottle. They *may* give you an extra day, but they just don't seem worth the effort and price to me. There is, however, a nifty contraption called the Coravin that lets you steal wine from the bottle through a surgical needle without ever pulling the cork. I don't think that counts since the bottle was never opened.

If you are determined to save leftover wine, there are a few tricks. The first is to reseal the bottle and put it in the fridge. It also helps if you pour the wine into a smaller bottle so there is less air in the bottle. I keep empty half bottles around for just that reason, but a plastic water bottle will work in a pinch. Just make sure you fill it to the very top and cap it.

Tale from the Wine Floor

Customer: How long will this jug wine last before I have to dump it out?

Me: I'd dump it out as soon as I opened it.

59. WHAT DO I DO IF THE CORK BREAKS?

Under the right conditions, *some* wines can better with age. That is not the case for cork. It most definitely gets worse. All cork will eventually dry out, break, and even crumble. That does not mean the cork didn't do its job of maintaining a seal. For the vast majority of bottles, the wine will still be fine to drink. So now what?

I have seen all kinds of hacks for opening a bottle of wine with a broken cork, including using a Swiss Army knife, a sheet metal screw, or even a shoe! But the best way to address this problem is to use an "Ah-So." This flat, two-pronged opener (known as a "Butler's Thief") takes a little getting used to, but once mastered, it may be your go-to opener for all of your wine bottles. To use the crafty opener, you place the longer prong between the cork and the glass and push in slightly. Then bend it back and insert the other prong. Now toggle them both down in a rocking motion. Holding the handle, you then rotate it up. Presto! The cork will come with it.

The simplest way to get to the juice under extreme conditions, however, may be to concede defeat and push the cork in. You can use anything that's handy and fits, like a teaspoon. When I'm at home, I like to use a knife honing steel. It is the right thickness, long enough, and even has a handle! ***Warning:*** Go slow. Pushing it in hard and fast will make a surprising splash. (I always put my hand over the opening.) Make sure the cork is completely dislodged and pushed past the shoulder of the bottle. Otherwise, if the partial cork dislodges while you are pouring, you'll pour half the bottle on the table.

If the cork has crumbled into the wine, a screen of some kind will be needed. I like to use a tea strainer. You can strain it directly into the glass or into a decanter.

The point is, it's not that big a deal. Annoying . . . but no big deal.

60. CAN I FREEZE A BOTTLE OF WINE?

The answer to this question is tricky. You can freeze wine, but not if it's in the bottle. Wine is mostly water and will freeze, but it will also expand and make the bottle crack. A better idea is to put the wine in ice trays. When you need a splash of wine for cooking, add a cube. *Never* try to freeze a bottle of sparkling wine. It will explode.

61. WHAT IS A "COOKED" WINE?

As you can guess, a "cooked" wine is wine that has been exposed to heat. It is not as easy to tell at first, but the wine smells sort of sweet but processed. It reminds me of stewed fruit or wine sauce. When the storage temperature is extremely high, the pressure inside the bottle will push the cork up ever so slightly. I have seen a cork push right through the capsule. Definitely cooked!

So, if a wine gets too hot, does it go bad? Not necessarily. Wine is more resilient than you think. The bigger problem is if the temperatures go up and down, which causes condensation to occur in the bottle. If a bottle is stored in hotter than optimal temperature (50°–55° F), it just ages faster—and, logically, if it's too cold, it ages slower. But if a wine *is* cooked, it can't be salvaged and isn't good for anything . . . not even cooking!

62. WHAT IS A "CORKED" WINE?

About 3% to 4% of wines—a bottle in every two cases—are known as "corked." The chemical compound that caused the problem only identified in 1981 by Swiss scientist Hans Tanner. 2,4,6-Trichloroanisole (or TCA) can make a wine smell or taste like musty, wet cardboard, even when it is found in infinitesimal amounts. Remarkably, some supertasters claim to detect TCA in ridiculously tiny amounts—between two to five parts per *trillion*! (If you're trying to wrap your head around that one, think of drops of water in an Olympic-size swimming pool, or 30 seconds in a million years.) How'd the chemical culprit get in there? Mostly from the cork, but it can also be present in the barrel and even inhabit and contaminate an entire winery! Is it harmful to drink a wine that is corked? No. Will it blow off if I decant it? Sometimes. Can a wine with a screw cap be corked? Oddly enough, yes. But it's less likely.

A few years ago I was at a tasting with a group of wine geeks and heard about a hack that takes away the smell and foul taste of a cork-tainted wine. Here's the reasoning and the process.

Supposedly, plastic wrap made with polyvinyl chloride (PVC) attracts and binds with the molecules that cause the problem and renders the wine TCA-free. As I investigated it further, I found that you can't use just any plastic wrap. The plastic wrap brands made with low-density polyethylene (LDPE) don't work as well. As of this writing, a brand sold at Costco called Stretch-Tite uses PVC. So, roll of plastic wrap in hand, I gave it a shot. I crumpled up a ball of the film, immersed it into the wine, and waited the required 10 minutes.

Does it work? Well, kinda. A significant amount of the cork taint was gone. Unfortunately, other components of the wine also disappeared. And I swear I could detect a slight taste of plastic on the finish. I guess I'll have to ask Hans what I can use to take that away. I hope he doesn't say "Cork!"

63. IS BAD WINE HARMFUL?

Wine can *be* bad or *go* bad for a multitude of reasons, but it's still alcohol—which is a natural preservative. The alcohol will protect the juice from developing bacteria that might actually do you harm.

There are a few tell-tale signs that will alert you that a wine has turned—it changes color. A white wine will get darker and a red wine will change to an unappealing cloudy brown. Next is the smell. A wine that is "corked," "cooked," or just past its prime will give off smells that may remind you of rotten eggs, vinegar, iodine, moldy cardboard, or a term that wine geeks love to use: wet dog. If you somehow miss the look and smell and get it to your mouth, your taste buds will let you know. The wine will literally be undrinkable . . . but not harmful.

Tale from the Wine Floor

Customer: Ya see this bottle here? I'm looking for the red version of it.

Me: That's a white grape. There isn't a red version of it.

C: Yeah, there is. I've had it before.

M: No disrespect, but that white grape is made only into a white wine.

C: But I'm positive I had the red version of this wine!

M: OK, well, in that case, we are out of it.

C: Do you know when you'll get more in?

64. WHY DO SOME WINE BOTTLES HAVE DIFFERENT-COLORED GLASS?

The function of the glass is not only to contain the wine but also to preserve the wine. Tinted colors block sunlight and preserve antioxidants that protect the wine from oxidation.

Bottles have different colors (and shapes), and each one has its own history. Traditionally, dark-green wine bottles are used for red wines, while amber bottles are usually for sweet white wines. Light-green wine bottles are typically for dry white wine and Champagne. Occasionally you even see blue bottles! That most certainly is marketing.

No matter the color, wine should be stored away from direct light. A wine, especially a white wine, exposed to light will turn its golden hue into a very unappealing shade of yellow.

Ultimately, the choice of color comes down to the winemaker's preference for tradition, wine integrity, aesthetics, or marketing.

Chapter Six

STORING

65. DOES WINE ACTUALLY GET BETTER WITH AGE?

It is estimated that 90% of wines are best consumed in the first year, and an amazing 99% within three to five years after that. Only a select few have the elements needed (tannins, acids, sugar, etc.) to age for extended periods of time. That is not necessarily a bad thing. I imagine that same small percentage describes people who prefer aged wine . . . or have a place to store them!

Winemakers are certainly able to make age-worthy wine. They can macerate the juice with the skins for longer periods of time. They can punch down and pump over the juice to extract every bit of the tannins from the grape's skins, seeds, and stems. They can manipulate the pH to accentuate acidity levels. Hell, they can simply add sugar—all the components needed for a wine to stand the test of time. But to what end? The modern-day consumer wants fruit-forward, uncomplicated wine. They like the primary taste of the grape and don't understand or appreciate the tertiary (secondary) flavors that *may* develop over time. So, winemakers, who also have mortgages, car payments, and college tuitions, make wines that consumers want to buy.

Back to our question at hand for a definitive answer.

When you consider the vast majority of wines that do not get better with age compared with the small percentage that do, you are left with the plain and simple answer: No. Wine does not get better with age!

Curious side note: There is a novel device called a *Clef du Vin* that claims to instantly age wine by simply dipping the tool into the glass. Supposedly, each second the mysterious "alloy" stays in contact with the wine mimics one year of aging. Wanna guess my opinion of it?

Fun fact: Eighty percent of wines bought in the United States are consumed within 48 hours.

66. WHY DO RED WINES AGE BETTER THAN WHITE WINES?

The main reason for aging a bottle of wine is to soften the tannins, which you know by this point in the book come from the skins, seeds, and stems. White wines don't have much tannin (unless it's an Orange Wine!) because white grapes get separated from their skins early in the winemaking process.

There are a few exceptions to the "white wine doesn't get better with age" rule. German Rieslings, Austrian Grüner Veltliner, Chardonnay from Burgundy, Chenin Blanc from the Loire, and dessert wines age incredibly well and take on tertiary flavors that weren't there in their youth.

But perhaps the ultimate exception to the rule is Château d'Yquem. The iconic and ultra-expensive Sauternes from the Graves region in France is possibly the world's greatest wine—red or white. The concentration and high acidity balance the sweetness of the wine to such an extent that, with proper storage, a bottle not only can last but also get more complex for a hundred years or more.

67. WHY IS CORK USED TO SEAL A WINE BOTTLE?

Cork is light, eco-friendly, elastic, and impermeable to liquid, but it will let about one milligram of oxygen into the bottle per year and thereby allow a wine with the right amount of tannins and acids to age gracefully. It is also one of the most amazing and mysterious products in nature. It comes from a tree that has a lifespan of over 150 years. The bark has two layers of cork, one living and one dead. As each successive inner layer dies, the outer layer becomes thicker, making it able to be stripped off over and over without damaging the tree. A single cork punched out of the bark is made up of 800 million 14-sided cells.

No one knows who the first person was to put a cork into a wine bottle, but most scholars (yes, there are scholars that study these things) believe the Romans (it's *always* those crazy Romans) were the first to seal wine containers with cork around 500 BC. The first corkscrew to be given an English patent went to clergyman Samuel Henshall in 1795.

Before cork was used, oil-soaked rags were stuffed into bottles. Cork was used without issue to seal virtually every bottle of wine for nearly 300 years, until 1981. That's when another mysterious element in cork known as TCA was found (as referenced in #62). There is no legal standard for acceptable TCA levels in wine since it doesn't pose any health concerns, but it does strip a wine of flavor in small amounts and makes it undrinkable in larger levels. There is considerable debate over the exact percentage of bottles ruined by cork taint, but whatever the number, the search continues to find the perfect alternative to cork.

68. WHICH IS BETTER: SCREW CAP OR CORK?

The cork versus cap argument will not be settled here or anytime in the near future. There are too many pluses and minuses to both closures, with proponents on each side of the debate that swear by their opinion. Here are a few pros and cons for each.

Cork is renewable, sustainable, allows for long-term aging, and has been a part of the wine tradition since the 1400s. However, it is also expensive, can break and crumble over time, and is sometimes prone to cork taint because of a mysterious chemical compound called Trichloroanisole or TCA. (See #67, "Why is cork used to seal a wine bottle?") The exact percentage is hard to know for certain, but it is estimated that cork ruins 3% to 4% of wine.

Screw caps, though patented in 1858, have been used for wine bottles only since the early 1960s. They are cheaper, easier to use, are less prone to TCA (but possible!), and they can be made to allow a percentage of air ingress for the aging process. They are also made of aluminum, which is not biodegradable. It is recyclable, but the plastic inner liners are not.

There are other closures on the market made from plant-based, polymer synthetic, and glass materials, and others probably in research and development right now. So, while I imagine a time when the screw cap is not associated with lower-quality wines, I do not think the traditional cork is going away anytime soon.

Australia was the first country to go all-in on screw caps. In 1973 an influential group of wineries, including Yalumba and Penfolds, started commercially releasing under the twist-off. By 2010, 80% of Australian reds were bottled similarly. Today, the vast majority (some estimates believe it to be 98%) of Australian wine bottles come with a screw cap.

Bottom line: The question has been around for 50 years without anyone coming up with the perfect way to seal a bottle of wine while still allowing it to age gracefully.

$\mathbf{T}$ale from the Wine Floor

Customer: Do you sell wine openers here?

Me: Certainly.

C: This bottle needs one, right?

M: No, that's a sparkling wine. It doesn't need a corkscrew.

C: So, it's a screw cap?

M: It's not really a screw cap, but it doesn't need a corkscrew.

C: Isn't it one or the other?

M: It's one or the other if it's a still wine but not if it's a sparkling wine.

C: Wine is so complicated. That's why I drink beer.

M: In a can or bottle?

69. WHY ARE WINE BOTTLES SUPPOSED TO BE STORED ON THEIR SIDES?

Cork is porous. Laying a bottle on its side keeps oxygen from getting into the bottle by keeping the liquid at the top. It also helps the cork to stay moist and prevents the cork from drying out. That is the consensus to date. *However*, with the ever-changing and ongoing science of wine closures, there is a debate as to its validity. A new theory contends that the small headspace of a sealed bottle provides 100% humidity, so there is no need to place bottles on their side to keep the cork damp—the cork won't dry out even if you store the bottle upright. Wine collectors think that's crazy talk. Statistically, most wine is drunk within weeks of purchase, so the question is moot. Wine racks do look cool, though.

70. WHY ARE WINE BOTTLES TYPICALLY 750 mL?

The exact and verifiable answer is lost to history, but there are theories to explain the peculiar size of wine bottles. Some contend that it is one-fifth of a gallon. Why one-fifth? No one knows for sure. Others claim it was the amount provided to a Roman soldier for daily consumption. But my favorite explanation is that the average glassblower's lung capacity making one continuous blow results in a bottle that holds 750 mL.

The wine bottle itself is relatively new to the world. For hundreds of years, wine was bought directly from the winery in containers supplied by the customer, each holding a different amount.

Until the 17th century, glass bottles were handcrafted luxury items made only for noble families and kings. Sir Kenelm Digby is considered the inventor of the modern wine bottle. He added his own secret ingredients to get the fire hotter, making the glass stronger and, more importantly, darker in color.

It wasn't until the 20th century that automation, rather than glassblowers, made bottles, and not until 1979 that the United States required wine bottles to contain 750 milliliters. In the national effort to change to the metric system, liquor is one of the few things America managed to switch. Other countries adopted the 750 mL standard to be compliant with and ship into the lucrative American market.

71. WHY ARE WINE BOTTLES PACKED 12 TO A CASE?

There is some logic to having bottles packed three-by-four in a box. It makes a rectangular case that's easy to stack in warehouses and stores. However, they've been packed that way long before there *were* warehouses and stores. Although there is no definitive answer, the dozen (French word *douzaine*, meaning "a group of 12") is one of the earliest primitive groupings, perhaps because the average human has 12 ribs. The number is used in astronomy and astrology because it divides without remainder into halves, thirds, and quarters.

A more simple explanation is that the Romans used to subdivide their money in groupings of 12—so calendars, clocks, eggs, and even Apostles all use 12. Hey, works for me.

72. DOES THE SIZE OF THE BOTTLE MATTER?

Most wine professionals believe that the larger the bottle, the slower the wine ages. The theory is that the ratio of liquid to air ingress is less in larger-format bottles. And less oxygen would allow the wine to age slower. Therefore, the opposite would also be true: The smaller the bottle, the faster it would age. The thickness of the glass also plays a part in a wine's lifespan. Larger bottles require thicker glass. So, size does matter!

Wine bottles in larger sizes have unique names inspired by Biblical kings and prophets. They are most often used by Champagne makers. Here they are, along with the equivalent to standard bottles.

- Piccolo (187 mL) = ¼ bottle
- Demi/Split (375 mL) = ½ bottle
- Standard (750 mL) = 1 bottle
- Magnum (1.5 L) = 2 bottles
- Jeroboam (3 L) = 4 bottles
- Rehoboam (4.5 L) = 6 bottles
- Methuselah (6 L) = 8 bottles
- Salmanazar (9 L) = 12 bottles
- Balthazar (12 L) = 16 bottles
- Nebuchadnezzar (15 L) = 20 bottles
- Melchior (18 L) = 24 bottles
- Solomon (20 L) = 26 bottles
- Sovereign (26 L) = 35 bottles
- Primat/Goliath (27 L) = 36 bottles
- Melchizedek/Midas (30 L) = 40 bottles

Fun fact: According to the Guinness Book of World Records, the biggest bottle ever sold was a bottle of Beringer Private Reserve Cabernet Sauvignon 2001, dubbed "The Maximus," holding 130 L (173 bottles). It was sold to a wine store in Tenafly, New Jersey, for $35,812. The smallest bottle was made by Pol Roger and named the Winston Churchill. It held exactly 20 ounces. Apparently, it was the perfect amount for Winston's morning routine.

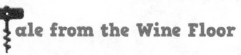

Tale from the Wine Floor

Customer: I'm looking for a wine. The only thing I remember about it is that it had a name from the Bible.

Me: Was it a word associated with the Bible, like Temptation or Redemption, which are Zins?

C: No, that's not it.

M: Was it Champagne? The bottle sizes have names from the Bible—Jeroboam, Methuselah, Balthazar.

C: No. It's the name of a person in the Bible.

M: How 'bout Dante, which is a Pinot Noir, or a Saint name like Saint Jean, St. Francis, or the region in Burgundy Saint-Veran?

C: Nope. Still not it.

M: Could it have been a kosher wine?

C: Wait . . . there it is! This Chianti is it.

M: DaVinci? Leonardo DaVinci wasn't in the Bible!

C: Where was he from?

M: He was from the Renaissance. A painter, sculptor, and inventor.

C: And not in the Bible?

M: No.

C: Are you sure?

M: I swear on the Bible.

73. WHY IS WINE MADE IN OAK BARRELS?

Barrels have been used to store and transport wine since the 1600s because they are easy to roll up the ramp of a ship and make turns. Winemakers now have other wine-storage choices (including stainless steel and concrete options), but over the convening two millennia, we've acquired a taste for oak and the other flavor components from the wood that get infused into the juice during the aging process after fermentation. Generally speaking, American oak imparts stronger flavors of vanilla and coconut, whereas European oak (most frequently French) offers textural subtlety with spicy notes. Both of these come at a price. With American barrels costing $700 and their French counterpart a whopping $1,200, you can safely estimate that two to four dollars paid for that wine bottle went towards the oak barrel. Another staggering fact is that an oak tree takes several decades to grow to maturity but produces wood for only *two* barrels. And the oak can be considered "new" oak only for three years. After that, it is deemed "neutral," since most of the seductive spice rack nuances loved by wine connoisseurs have been depleted. Technology, as always, has come to the rescue in the form of additives that mimic the taste of oak. The good news is that the lifespan of the barrel doesn't end with wine. Enterprising wineries sell their used barrels to whisky producers, who then sell them to beer makers, Sherry makers, Scotch makers, and Bourbon makers, who then sell them *back* to winemakers! Yes, there are now quite a few wines that tout Bourbon Barrel aging for their Zins, Cabs, and even Chardonnay.

At the end of a barrel's life, it can be separated into individual staves or even ground into usable oak sawdust.

Fun Facts About Wine Barrels

- A typical wine barrel holds 60 gallons (300 bottles/1,500 glasses).
  - There are an estimated 5.2 billion white oak trees in the United States, covering approximately 235,000 square miles.

- ○ Anyone with the surname Hooper or Cooper probably has ancestors who worked the ancient craft of "Cooperage."
- There are no knots in the staves of an oak wine barrel.
  - ○ A barrel that starts with white wine can only be used for white wine; the same is true for red wine.
  - ○ Barrels are charred on the inside to give more layers of flavor. The toast levels are light, medium, medium-plus, or heavy. But the original char comes from the heating process used to bend the staves.
- The 2% to 5% volume lost in the barrel to evaporation is called "Angel's Share."
  - ○ The largest barrel, built in 1751, used 130 oak trees and is housed in a German castle. It stands 7 meters high and 8½ meters wide and holds 58 thousand gallons.

Tale from the Wine Floor

Customer: Can you help me find a wine I can't remember the name of?

Me: Sure.

C: It begins with a K.

M: Is it Kendall-Jackson?

C: I don't think so. It had a dash in the name.

M: Kendall-Jackson has a dash.

C: No, the guy who makes it was on that TV show about bosses.

M: That's Kendall-Jackson.

C: Let me call my wife. She'll know.

(Calling) "It's me. What was that wine we wanted to get . . .
you know, the one from that show we watched? Kendall-Jackson?
OK, great."

It's called Kendall-Jackson.

M: Uh-huh.

74. HOW DO I START A WINE COLLECTION?

To my way of thinking, there are only two reasons to start a wine collection: convenience or investment.

A convenience collector simply wants to have enough choices for dinner, a BYOB, or a gift for friends without having to run to the nearest liquor store every time. They typically buy onesies and twosies from anywhere and everywhere without any thought other than having enough bottles they know and like to suit their needs.

An investment collector has a "buy low, sell high" mindset. They will study vintage charts and auction lists with the hope of buying a case or two of the highest-rated wine from regions that have aging potential, track the wine's progress by tasting over time, and then selling.

Some wine lovers may not even realize they've started a collection until they have one.

For our purposes, let's forgo the idea of building wealth through wine and instead focus on when and how to build a collection for the everyday wine lover . . . albeit one with a little discretionary income.

The first thing to consider is storage space. You can get pretty creative these days. If you're handy, there are DIY kits that can help you convert almost any space into a wine "cellar." If, on the other hand, you have grand plans and a budget to match, there are companies that can design and install your wine room vision.

The next thing to consider is temperature (45° to 64° F). Keeping the wine at a constant temperature is what is most important. An "active" cellar uses a wine-specific cooling system to maintain the proper temperature and humidity. In contrast, a "passive" cellar is usually built in a space that maintains both temperature and humidity naturally without the help of an external unit. This is obviously ideal since it costs nothing to operate and is not susceptible to power outages.

Now comes the fun part: Buying. I like the "now or later" approach. For now: These are your go-to wines that you drink regularly. They aren't rare or too expensive, so you won't get upset if a family member opened one without asking first. For later: These bottles have the potential to age but can be enjoyed without guilt if the right occasion presents itself.

Even the most modest collectors may have a few jewels that they just can't imagine *ever* opening, which is a shame. Estate buyers or your children might not share your passion or know how rare that first-growth Bordeaux is. Be careful not to treat your wine like it's a family heirloom.

Get to know your "wine guy" (or girl) at a reputable store.* He or she should be able to point you to special bottles and buys. Wine-club mailing lists and online retailers are also good places to grow your collection.

Just like the wine itself, balance in a wine collection is important. Keep a mixture of varietals, regions, and vintages.

Lastly, whatever the perfect number of bottles you have in mind, double it!

Note: A "reputable" store should have a comfortable layout and lighting that makes it easy to browse, have a decent selection of current vintages from most major countries, possibly a separate, temperature-controlled storage room for rare bottles, and a knowledgeable and friendly staff. It's a red flag if the store's temperature is uncomfortably hot or bottles are on display in direct sunlight.

Chapter Seven

NAMING

ale from the Wine Floor

Customer: Can you tell me where Menudo is?

Me: The boy band? I think Puerto Rico.

C: No, the wine Menudo.

M: I think the wine Menudo you're thinking about is Moscato. Menudo was the band Ricky Martin was in.

C: Are you sure?

M: About Ricky Martin? Yeah, I'm sure.

C: No, about Moscato.

M: Yeah, I'm sure about that too.

75. WHAT IS AN APPELLATION?

Often, one of the first things you ask when meeting someone new is "Where are you from?" It's an easy way to begin the process of getting to know the person. It's also how we define ourselves. Where you were raised had an enormous impact on shaping who you are today. The same can be said about wine grapes. A grape's hometown is called an appellation, and, in the United States, the hometown is called an AVA (American Viticultural Area). Some even have sub-appellations within their appellations, like Russian nesting dolls.

Finding that perfect place to grow that perfect grape takes a lot of trial and error. But it's not just wine that is so site-specific. Look at it this way: A Jersey tomato is better than one grown in, say, Idaho. Then again, Idaho grows a better potato than Florida, which grows better oranges. Now, before we take it to the extreme and argue who makes the best pizza (Taconelli's), let's just say that the place matters no matter what the product.

During the winemaking process, governmental supervision steps in to ensure authenticity and prevent fraud on behalf of consumers. Without this oversight, unscrupulous merchants might say they are bottling something they are not. A Napa Cab must be made with grapes grown in Napa, California. (OK, not 100% of the grapes, but that's another story.)

Although appellations were mentioned in the Bible (the wine of Jezreel), the first officially sanctioned vineyard was in Chianti, Italy, in 1716. However, France was the first country to geographically map out wine regions and set the original standards to which winemakers must adhere. Their AOC (Appellation d'origine contrôlée) model from 1935 has been copied around the wine world. Some of their appellations have become so familiar they have become a generic term, a process known as genericization. Scotch Tape, Kleenex, and Google are examples of trademarks that have become generic terms. Much to France's chagrin, most of the best-known, and thus the most genericized, wines are from France. If you refer to everything that has bubbles as "Champagne," then you're guilty as charged.

In the United States, the Alcohol and Tobacco Tax and Trade Bureau (TTB), a division of the Treasury Department, regulates the geographic boundaries of appellations. But American grape growers have much more leeway than their French counterparts. They are free to plant any grape varieties they wish and harvest as large a crop as possible.

In France the INAO (Institut National des Appellations d'Origine) enforces rules that not only approve the grape variety grown but set the area boundaries, the yield, the ripeness levels, and the maximum alcohol levels. They even regulate the pruning methods and allowable vineyard spacing!

As each country adds more and more appellations, it is becoming a veritable alphabet soup of wine. To make matters worse, the European Union has added PDOs (Protected Designation of Origin) to "simplify" things. I understand the adage *location, location, location,* but all these AOCs, DOCGs, DOs, GIs, and QBAs are making me wanna grab a VSOP and take a NAP!

Fun fact: The largest AVA is the Mississippi River Valley. It stretches across four states and encompasses over 29 square miles.

ale From the Wine Floor

Customer: Do you have a wine called "Vino"?

Me: Well, actually all wine is vino. Vino means wine.

C: I know, that's what I'm looking for . . . the wine.

M: Which?

C: The one called Vino.

76. WHY DO THE FRENCH NAME WINES AFTER PLACES?

Well, now you've done it! You've come to the point of no return, also known as French wines. As daunting and complicated as it seems, I'm here to tell you—it is! Let's jump in anyway.

The French take their national drink seriously. I mean *really* seriously. They have laws that regulate everything. So, the key to understanding French wines is to have a little understanding of the regions of France. Some regions have smaller regions within them, but there's no need to get into the weeds. Knowing a few main regions and the grapes that grow there will keep you busy enough. You can wander in all directions after that.

For starters, the French are not trying to capture the essence of a *grape*; they are trying to capture the essence of a *place*. That sense of place is what the French call "terroir," which can be loosely defined as the combination of grapes, soil, climate, vineyard placement, and even the human influence that goes into making a wine. Producers believe that where the grape is grown is more of an indication of the taste than merely stating the grape itself. That is essentially why French wines list the place on the label and not the grape. This is confusing for many, but for the French, it is a nonissue. After making and drinking wine for thousands of years, the French just *know* which grape is grown in which appellation. Once wine became an international business, the French didn't bow to the system of naming wines after grapes that seemingly everyone else was using. Instead, they stuck to their tradition of naming wines after French places. As a somm, I respect their decision. As the saying goes, if it ain't broke, don't fix it.

The rules that French winemakers must adhere to are known as the AOC laws. AOC stands for *Appellation d'origine contrôlée*, or controlled designation of origin. Though these rules initially sound oppressive, the laws preserve the geographical origin, guarantee the quality, and indicate the style of the wine. Generally speaking, the more specific the vineyard, the better the wine. Rather than go down the rabbit hole of the 450 AOCs in France, let's focus on the five that are most known.

Champagne

Champagne is a place. If the sparkling wine is not made in that place, it's not Champagne. Now, there are a lot of "sparkling wines," and some are very good, but the process of making a "real" Champagne differs from them. The Champagne process, or *méthode champenoise*, involves a secondary fermentation that happens *inside* the bottle. The main thing you need to know is that it is usually a blend. The three grapes used are Chardonnay, Pinot Noir, and Pinot Meunier. If the label says Blanc de Blanc, it's 100% Chardonnay, and if it says Blanc de Noir, it's 100% Pinot Noir. Most Champagne is "non-vintage," meaning the grapes came from multiple years. If the conditions are perfect, they will make a "vintage" Champagne composed 100% of grapes grown in that year. Lastly, there are varying sweetness levels to consider. (See #85 for the list of sweetness levels.)

Bordeaux

While there are red and white Bordeaux, most people associate Bordeaux with red wine. Wines from Bordeaux are blends and are often named for the chateau where they are produced. By the way, "chateau" doesn't always mean a sprawling stone mansion. More often than not, it's just a house.

The main two grapes in red Bordeaux wines are Cabernet and Merlot, but winemakers may also add Cabernet Franc, Petit Verdot, and Malbec (Carménère is allowed but rarely used). White Bordeaux uses Sauvignon Blanc, sometimes with the addition of Sémillon and/or Muscadelle. The region is divided by the Gironde River. The Left Bank is dominated by Cabernet Sauvignon blends, while the Right Bank is best known for Cabernet Franc and Merlot blends. The Gironde then splits into two other rivers—the Garonne and Dordogne—into what looks sorta like a peace sign. Groovy.

One of the key reasons Bordeaux is so prized is due to its aging potential. Time allows the tannins in the wine to soften and the different components of the grapes to further blend together, resulting in the complexity that a good percentage of wine drinkers crave. Other wine drinkers (like me) love that classic blend at a young age. (See #65, "Does wine actually get better with age?")

Note: During the writing of this book, new grapes have been approved for inclusion in Bordeaux wines as a direct result of climate change. Depending on the region, the seven new allowable grapes are:

- **Touriga Nacional**–A red grape best known and grown in Portugal's Douro Valley.
- **Alvarinho**–A white grape from Portugal and Spain.
- **Marselan**–A cross between Cabernet Sauvignon and Grenache that was created in a French research institute in 1961.
- **Petit Manseng**–A white grape usually used in sweet wines.
- **Arinarnoa**–A cross between Cabernet Sauvignon and Tannat.
- **Castets**–A rare red variety first identified in Bordeaux in 1870.
- **Liliorila**–A white variety that is a cross between Chardonnay and an obscure grape with the coolest name, Baroque.

Keep in mind that you will not see these varietal names on a Bordeaux bottle. Then again, you rarely saw the previous grapes listed either. Growers will be allowed to plant up to 5% of them in their vineyards and add up to 10% in their final blend.

Burgundy

In contrast to Bordeaux, Burgundies are all single grape wines (no blends). Although there are few other grapes grown in the region (Aligoté is one of those grapes that is becoming quite trendy of late), it is safe to say that a red Burgundy is 100% Pinot Noir and a white Burgundy is 100% Chardonnay. Actually, that is all you need to know when you walk down the Burgundy aisle, but here's a little more info. Wines made in Burgundy receive their quality classifications based on the region, village, or vineyard rather than the "chateau" as seen in Bordeaux. The four quality tiers in descending order are Grand Cru, Premier Cru, Village wines, and Regional wines. Not surprisingly, Burgundy prices match their classification, (i.e., Grand Cru requires a "grand" budget, and so on).

As in Bordeaux, the region is exploring measures to moderate the effects of global warming on their grapes. In the very near future, we may see other clones, varietals, and possibly other species of grapes allowed in Burgundy.

Note: Beaujolais is technically located in Burgundy but uses its own classification system.

Côtes du Rhône

Without a doubt, the region that offers the most consistent quality and best value in France is the Rhone Valley. With 100,000 acres of vineyards owned by more than 10,000 growers, it is the second-largest appellation in the country. (Bordeaux is the first.) In addition to the value of Côtes du Rhône wines, there is another advantage: With a few exceptions, they are meant to be enjoyed upon release. They can be red, white, or Rosé wines and are generally known for blends dominated by Syrah and Grenache, although a large number of varieties are allowed in the AOC. The iconic Rhône Châteauneuf Du Pape can use as many as 13 different red and white grapes.

Loire

If there was ever a region that produced wines that can be simply described as "delicious," it's the Loire Valley. France's longest wine region stretches 630 miles along the Loire River and is divided into four major regions, the Upper, Middle, Center, and Lower Loire. Within each region are over 80 appellations! This pastoral le jardin de la France (Garden of France) and UNESCO World Heritage Site is renowned for the quality of four main grapes: Sauvignon Blanc, Chenin Blanc, Melon de Bourgogne (Muscadet), and Cabernet Franc, although Pinot Noir and Gamay are also grown. This is not the land of big and bold, high-alcohol wines. Light, fresh, and affordable are the hallmarks of the Loire Valley.

Sacrebleu! That was a long answer. I need a glass of French wine after that one.

ale from the Wine Floor

Customer: I'm buying a wine for a friend, but I only know the first name of the wine he drinks.

Me: What's the name?

C: Domaine.

M: I think I'll need more to go on. Ya see, Domaine translates to "estate," or property. It's mainly used in Burgundy. It would be like asking for a person but only knowing that his first name was Mister.

C: Does Barefoot make a Domaine?

M: No.

C: OK, I guess I'll get him a Domaine Burgundy then.

ale from the Wine Floor

Me: Can I help you find anything?

Customer: Yes, Chablis.

M: Real Chablis or the jug wine called Chablis?

C: Aren't they the same?

M: No. Real Chablis is from Chablis, France, and is made using Chardonnay. The jug wine version is from who knows where and uses who knows what, but I doubt any Chardonnay.

C: So, does real Chablis come in a jug?

M: No.

C: How 'bout Burgundy . . . the hearty one.

M: Here we go again.

C: Don't tell me. There isn't Burgundy in Burgundy either?

M: Well, Burgundy is also a place, and the grape is Pinot Noir.

C: OK. I want what *used* to be Chablis and *used* to be Burgundy. In a jug.

77. WHAT'S THE DIFFERENCE BETWEEN BEAUJOLAIS AND BEAUJOLAIS NOUVEAU?

Beaujolais is a region in Burgundy, France, and Gamay is the grape. Beaujolais Nouveau (translation: new) uses juice that is vinified quickly and made in a drink-now style. Historically, Nouveau, with its quick fermentation, was considered a celebratory wine for the workers at the end of harvest. To get this unique expression of the grape, a unique method called carbonic maceration is used. During the process, whole grapes are put into a sealed container that is then filled with carbon dioxide. The grapes at the bottom are gently crushed by the grapes at the top and undergo fermentation without extracting bitter tannins from the grape skins. With no tannins, and therefore no aging ability, it is recommended that you drink it within months of release. Beaujolais Nouveau is a red wine that drinks like a white wine. You can chill it or even drink it over ice!

However, the story of Beaujolais Nouveau is much more interesting than the wine. The Gamay grape dates back to the 14th century, when Philip the Bold, Duke of Burgundy, ordered most of the grapes to be replaced with his favorite Pinot Noir. The grapes that survived made simple wine for chugging by the locals for hundreds of years. When the first AOC laws were written in the 1930s (see #75, "What is an appellation?"), it was ordered that Beaujolais Nouveau couldn't be sold until after December 15 in the year the grapes were harvested. Then, in 1951 the rules were relaxed, allowing it to be released on November 13–a date that coincided with America's Thanksgiving. And what is the perfect grape that pairs with Thanksgiving turkey? Gamay!

Beaujolais Nouveau gained more fame with the help of negociant Georges Duboeuf. (A negociant–pronounced "nay-GO-see-aunt"–is a merchant who buys grapes, juice, or finished wine from growers.) He is credited with marketing each new wine release with a race from Beaujolais to Paris that used the catchphrase "Le Beaujolais Nouveau est arrivé!" ("The new Beaujolais has arrived!"). The wine race enjoyed the 1960s version of "going viral" when Allan Hall, a British newspaper columnist for the *Sunday Times*, offered a reward (a swap for a

bottle of Champagne!) to the first person who could get that year's new Beaujolais to his London office . . . a 500-mile journey. The media sensation expanded worldwide when the newly invented supersonic airliner known as the Concord was utilized for the race to New York, where the swap was made.

In terms of quality, next up the Beaujolais ladder is the "Village" level. There are 38 village levels, 30 of which can add their name to the label. Not much of a story here: They're just enjoyable wines that are reasonably priced and can please both newbie wine drinkers and pretentious wine geeks.

Finally, we come to a favorite of mine, "Cru" Beaujolais. *Cru* is a French term meaning "growth." However, in wine terms, it relates to a superior growing site or vineyard. Wines from these 10 Crus can compete with many fine Burgundies at a fraction of the cost. Each Cru has its own character and flavor profile. From a ratings standpoint, the recognized best are Moulin-a-Vent and Morgon, but each has its own charm. All are complex wines with alluring aromas, spice notes, and aging potential. What more can you ask of any wine?

Helpful hint: How do you tell the difference when you see Beaujolais in the French aisle? A Cru or Beaujolais-Villages won't say "Nouveau" on the label. Here is a quick snapshot of the 10 Crus:

1. **Moulin-à-Vent**—Named after a windmill that still stands on the property, this is the odds-on favorite for the Beaujolais that can best stand the test of time. If there were such a thing (and there is not), this would be the King of Cru Beaujolais.
2. **Morgon**—Second only to Moulin-à-Vent, this full-bodied Cru can go toe-to-toe with quite a few higher-priced Pinots from Burgundy. Look for wines coming from Côte de Py.
3. **Saint-Amour**—Soft, fruity, and floral. It can be lighter or bolder in style, depending on the producer.
4. **Juliénas**—Richer and more complex than most Crus, it has distinctive notes of vanilla and cinnamon.
5. **Chiroubles**—Lively and fresh with high acidity due to its high elevation, this is the most Beaujolais-like of all Beaujolais!
6. **Fleurie**—Appropriately named and most alluring, Fleurie is silky and refined.
7. **Côte de Brouilly**—Ripe and rich with a slight pepper edge, this is possibly the most unique Cru.

8. **Brouilly**–Coming from the largest vineyard of the Crus, it is medium-bodied with surprising black fruit notes. Best when young.

9. **Régnié**–Newest of the Crus (elevated in 1988), it is aromatic and more structured than most Crus.

10. **Chénas**–Smallest of the Crus in vineyard size, it is full-bodied and spicy.

If the only Gamay you've ever tasted was in a Beaujolais Nouveau, do yourself a favor and try a Beaujolais-Villages or one of the Cru Beaujolais. And don't wait until Thanksgiving!

Tale from the Wine Floor

Customer: Can you pick out a wine for me?

Me: I'd be happy to. What kind of wine do you like?

C: I don't know.

M: OK, do you like dry or sweet wine?

C: Hmmm . . . don't know.

M: Red or white?

C: Umm . . . I don't know. If you were me, what wine would I like?

M: Well, if I were you I'd like a wine from France—like this one.

C: Do you think I'll like it?

M: If you were me, you would.

78. WHAT IS A MERITAGE?

A Meritage is a Bordeaux-inspired blend that is not made in Bordeaux. In 1989 Agustin Huneeus of Franciscan Winery, Mitch Cosentino of Cosentino Winery, and Julie Garvey of Flora Springs wanted to find a suitable name for their American wine blend using classic Bordeaux grapes, so they had a contest. By combining the words "merit" and "heritage," Neil Edgar submitted the name "Meritage" that was chosen from over 6,000 entries. It is pronounced like "heritage" with an "m," not Meri-TAJ. The first Meritage was produced by Mitch Cosentino: the 1986 vintage of "The Poet."

The red varietals used in Meritage are Cabernet Sauvignon, Merlot, Cabernet Franc, Malbec, and Petit Verdot. The white varietals used are Sauvignon Blanc, Sémillon, and Muscadelle. No single grape can make up more than 90% of the blend. To legally use the word Meritage on the label, a winery must be a member of the Meritage Alliance. They own the trademark.

Note: Claret is a generic term for Bordeaux that is sometimes used by the British. There are a few American wine producers that use the term to describe their Bordeaux blend. Unlike Meritage, Claret is not trademarked.

79. WHAT'S THE THING WITH ITALIAN WINE?

If I had to explain Italian wines to say, one of the Sopranos, it might sound something like 'dis.

Yo, here's the thing.

What thing?

The thing about Italian wines. That thing.

Oh yeah, that thing.

To start wit, they got like 7,000 different grapes and make about a fifth of the world's wine.

Are they all good?

Some good, some not so good–but mostly good.

So, what's the problem?

The problem is trying to make rules that keep everybody happy.

Can't they make them an offer they can't refuse?

Stunod, they got like 900,000 vineyards. Fuhgetaboutit!

In all seriousness, Italian wines can be confusing. France and most of Europe name the wine region and not the grape. America and the New World name the grape and not the region. Italy does both. On top of that, they have names that *could* be the grape and *could* be the region.

Barbera is a grape that comes from Piedmont. Barolo is also from Piedmont, but that's a village, not a grape. There's a wine called Montepulciano d'Abruzzo made from the grape Montepulciano in the Abruzzo region. And there's also Vino Nobile di Montepulciano made from the grape Sangiovese, which is made in Tuscany. Wait, it gets worse. Amarone isn't a grape or a region–it's a winemaking process.

There are four official classifications, three subcategories, and one that has no official definition–not that everybody follows them. These wine classifications are kind of like traffic lanes in Italy–they're just suggestions. If that weren't enough, the European Union changed the classification letters in 2008, but most winemakers don't use them. Here's what these classification letters are meant to be:

1. Vino da Tavola (VDT/Table wine)—The winemaker of a VDT wine has little restriction as to the grapes he uses and doesn't even need to state the grapes on the label. Sometimes you just want a basic wine—you don't want to spend a lot, and you aren't expecting a lot. That everyday bottle of wine is Vino da Tavola.
2. IGT (Indicazione Geografica Tipica)—IGT was established in 1992. The wines are required to list the area where the grapes were grown but allow the winemaker the freedom to use non-traditional blends.
3. DOC (Denominazione di Origine Controllata)—DOC wines can come only from specific areas and use only specific grapes. Established in 1963, it regulates the amount of grapes grown, the blends, aging, alcohol levels, and the location of the vineyards and cellars. There are about 300 wineries within the DOC classification.
4. DOCG (Denominazione di Origine Controllata e Garantita)—DOCG wines are the top tier of the four levels and reserved for Italy's highest-quality wines. It was introduced in 1982. That letter G means a governing body Guarantees that the wines are taste-tested and have undergone chemical analysis before being bottled. There are currently 74 DOCGs:
   • Classico—The grapes come from the original or "classic" section of the region.
   • Superiore—They can be of a higher quality, but it generally means they have a higher alcohol level.
   • Riserva—The wine has spent a longer time in oak and is their best, in the winemaker's opinion.

Tale from the Wine Floor

Customer: Can you help me find a wine?

Me: I'd be happy to.

C: I don't remember the name, but it said "Italy" on it.

M: Well, here is our Italian aisle. Can you remember anything else about it?

C: Nope, that's all I know. Maybe it's in another aisle. Does France make Italian wine?

M: No.

C: What kind of wine do they make?

M: French wine.

C: So who else makes Italian wine?

M: No one else. All Italian wines come from Italy. There are Italian varietals that are made in other countries.

C: Do they say Italy on it?

M: No.

C: Well how am I supposed to find the wine I want if they don't say what it is?

M: I was just thinking the exact same thing.

80. WHAT IS A "SUPER TUSCAN" WINE?

A Super Tuscan wine is typically of a higher quality but uses traditionally unsanctioned grapes in the blend. Winemakers were frustrated that their creations would have to be relegated to the lower-tier VDT "table wine" category. The government was also unhappy that the appropriate taxes couldn't be applied. The problem was resolved in 1992 when the IGT (Indicazione Geografica Tipica) designation was added.

Piero Antinori's Tignanello, with its blend of Sangiovese, Cabernet Sauvignon, and Cabernet Franc, is generally regarded as the first Super Tuscan wine. But the real renegade was Mario Incisa della Rocchetta, who bucked the system by planting Cabernet Sauvignon at his Bolgheri Tenudo San Guido estate back in 1944. That estate is better known today as Sassicaia.

Although sometimes attributed to wine writer Burt Anderson and other times to wine critic Robert Parker Jr., no one is really sure who coined the American-sounding term *Super* Tuscan.

Tale from the Wine Floor

Customer: Is this wine the same guy who did *The Godfather*?

Me: Yes. Coppola is Francis Ford Coppola. Same guy.

C: Do you like it?

M: I do. Coppola makes some nice wines.

C: I mean the movie.

M: Oh. Yeah, I liked that movie.

C: How 'bout *Goodfellas*?

M: Yeah, I liked that too.

C: Does that guy make a wine?

M: Martin Scorsese? I don't think so.

C: He should, don'tcha think?

M: I guess.

C: So, what would you recommend?

M: Wine or movie?

81. HOW DO YOU SAY THAT?

Wine is grown in every country in the world. So it goes without saying that there will be different names for grapes and regions. Some are real tongue twisters. I dare say, some wines are not ordered in a restaurant because the person ordering couldn't pronounce the name of the wine! Thank goodness most lists have numbers. Am I right?

Here are the top 10 most mispronounced wines with a key to their pronunciation.

Who would like to go first? Say it loud enough for the whole class to hear.

Côtes du Rhône–(Coat-du-ROHN) No S sound.
Pouilly-Fuissé–(Poo-YEE-fwee-SAY) No L sound.
Pinot Gris–(Peeno-GREE). No S sound.
Semillon–(SEH-me-yhon) No L sound.
Viognier–(VEE-oh-nyay) No G sound.
Gewürztraminer–(Guh-VERTS-tra-mee-ner) No W sound.
Graves–(Grahv) No S sound.
Meursault–(Merr-SO) No T sound.
Tempranillo–(Tem-prah-NEE-yoh) No L sound.
Côte de Nuits–(Coat-deh-N'WEE) No T sound.

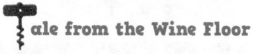

Tale from the Wine Floor

After years of fielding wine requests, I've learned to translate what people mean.

Here's my key.

Shutter House = Sutter Home

David Morgan = Mogen David

Charlotte Russo = Carlo Rossi

Manachevie = Manischewitz

Prosciutto = Prosecco

Barefeet = Barefoot

Stick wine = Zaccagnini

Kangaroo wine = Yellow tail

Rooster wine = Rex-Goliath

Grape wine = Arbor Mist

Blackberry wine = Wild Vines

Lani da Lana = Luna di Luna

Gnarly Toast = Gnarly Head or Toasted Head

Seven Crimes =19 Crimes or Seven Daughters

Crunchy Bull = Concha Y Toro

Khvachkarbaykapechta = Any Russian wine

Chapter Eight

FORTIFYING AND SPARKLING

82. WHAT IS FORTIFIED WINE?

In short, fortified wine is a wine to which a distilled spirit is added. The long answer is, before refrigeration was invented, wine was transported by ship in casks that were not airtight. As a result, the wine would oxidize and turn into acetic acid (vinegar). To prevent this, winemakers added alcohol to the wine, which resulted in less spoilage. The fortifying liquor is called a "neutral grape spirit"—essentially a brandy. Interestingly, there was a time in the not too distant past when the most planted wine grape on the planet was the Airén grape due to its use in fortifying wine. The workhorse grape native to Spain has slipped in recent years but still maintains an astonishing fourth place.

The amount of time a wine is allowed to ferment before being fortified determines whether it will be sweet or dry. Once the spirit is added, the yeast stops converting sugar to alcohol, and the remaining grape sugar is left in the wine. Winemakers can control how sweet or dry their fortified wine is by adding the spirit at different times during the process. The most common types of fortified wines are Porto, Sherry, Madeira, Marsala, and Vermouth. However, fortified wine means any wine between 16% and 24% alcohol by volume.

Madeira is a fortified white wine from the Portuguese island of the same name with a great tradition and history. It was the drink of choice of our Founding Fathers, who raised a glass of Madeira to toast the signing of the Declaration of Independence! Unfortunately relegated to the kitchen these days, they are powerful wines with a unique bouquet and complexity. They are made in a wide range of sweetness—including Dry, Medium Dry, Medium Sweet, or Sweet. The best is from a single harvest and aged for up to 20 years in cask.

Since Madeira is heated and oxidized, it pretty much can't be beat up any further. Think of it as a wine that has gone bad but tastes good!

Marsala is produced in the region surrounding the city of Marsala in Sicily and uses a process called *in perpetuum*, which is similar to fractional blending in the solera system. (See #84, "How is Sherry made?") The main white grapes used are *Grillo* and *Insolia*. Marsala is unique among fortified wines in that it indicates the sugar content on the bottle. Although there are five styles of aged Marsala available, the main two you'll see are a young Fine (aged for a minimum of four months) and an older Superiore (aged a minimum of two years).

Vermouth is an "aromatized" wine, meaning various botanicals (roots, barks, flowers, seeds, herbs, eye of newt, etc.) are added. It was produced in the mid-18th century in Turin, Italy, and was originally used for medicinal purposes. The name comes from *Wermut*, the German word for wormwood, an ingredient historically used in the drink. Although there is a whole range of styles available, the two main types are simply red (sweet) and white (dry). Vermouth can be sipped on the rocks as a light aperitif, but it also works in a wide range of classic cocktails from Manhattans to Martinis to Negronis. While inexpensive brands exist, craft vermouth makes a very big difference and is worth the search and the few extra dollars.

83. WHAT'S THE DEAL WITH PORT?

Port is a wine originating from the Douro River valley in Portugal and is named after the town of Oporto. The wine is a blend of grapes, the most common being Touriga Nacional, Touriga Franca, and Tinta Roriz (a.k.a. Aragónez and Tempranillo). It is also fortified, meaning the neutral spirit Aguardente (translation: "Firewater") is added. This stops the remaining sugar from fermenting and raises the alcohol content to about 20%. Similar to the rules for Champagne and Cognac, if it is not from Portugal, it is technically not Port. Port from Portugal will have the word *Porto* on the front label.

Interestingly, the British are most responsible for the invention of Port. During a political conflict in the late 17th century, the English boycotted French wine and replaced their love of Bordeaux with red wine from Portugal. To protect the juice from spoiling during the long voyage back to England, they added a "wee bit" of brandy. Various Port styles exist depending on oak-aging, filtering, time in the barrel, or time in the bottle. Here are the various styles:

Ruby Port: So named for its distinct dark garnet color, Ruby Ports are young, approachable wines with fresh, fruit-filled aromas. It is a blend of a number of grapes from one or more vintages and matured for a minimum of three years in oak casks.

Tawny Port: Tawny Port is matured for a longer time in barrels, typically 5 to 7 years, but it can be aged for 40+ years. Tawny Port takes on more of a nutty character with less overt fruit than Ruby Port, but with greater aroma and complexities resulting from its controlled micro-oxidation through the barrel. Tawny Ports come in three different styles: Colheita, Crusted, or Indicated Age. Colheita Port is made from grapes that all share the same vintage year. Crusted Ports are a fairly new invention. They are blended from two or three harvests and bottled unfiltered, which leaves sediment (crust) to form with time. Decanting may be needed before serving. Tawny Port may be bottled with the permitted age indications of 10, 20, 30, 40, and over 40 years.

Late-Bottled Vintage (LBV): An LBV Port is a Ruby Port made from grapes of a single harvest—the vintage that can be stated on the label—and can be filtered or unfiltered. They are bottled after aging in barrel for 4 to 6 years. Filtered LBV Port is popular with consumers because of its vintage style without the hefty price tag.

Vintage Port: Vintage Port is produced from a blend of the best grapes in a single vintage worthy of the highest-quality rating. It is bottled without any filtering following a short two-year maturation period in oak casks. It can then be aged in the bottle for 10, 20, 30, or more years. The structure of the wine and its aging potential are the major criteria for a Vintage Port. If all quality factors align, Portugal's Port Wine Institute approves it and declares it a Vintage year. There have been only 35 declared Vintage Ports since 1900.

Rosé Port: Rosé Port is a recent addition to the market. It is technically a Ruby Port but made like a Rosé wine, in which a little exposure to the grape skins gives it a rose color. It can be served over ice or used as a mixer.

White Port: White Port is made using the region's allowable white grapes in a wide variety of styles, from dry to very sweet, though typically drier than red. They can be bottled with the same age indications as Tawny Ports or as a Colheita. There is a separate subcategory of "Light White Port" with a minimum alcohol level of 16.5%.

84. HOW IS SHERRY MADE?

Sherry is made using a complex system of maturation called "The Solera Method," which I will explain in a moment. But first I feel it is my obligation to let you know that Sherry is perhaps the wine world's best-kept secret, which is odd because it was hugely popular throughout history. At one point, it was regarded as the world's finest wine. Shakespeare wrote about it in many works; Columbus took it with him on his second trip to America; and Magellan spent more money on it than on weapons. Have I piqued your interest yet?

Sherry gets even more interesting when you learn how it's made. The adage "great wine is made in the vineyards" does not apply to Sherry. Great Sherry is made in the cellars, and the cellars must be in Spain—specifically, in regions regulated by the Denomination of Origin (DO) system. Over 90% of Sherry is made from a low-acid grape called Palomino, but Muscatel and Pedro Ximénez also can be used.

First, wine is made by fermentation of the grapes. Barrels are then filled to about three-quarters to allow for a milky layer of yeast called "flor" to form on the surface. Under the veil of flor, the ever-evolving juice constantly interacts with the wine as yeast cells consume compounds and create others. The resulting neutral wine is then ready for the fun part.

(Drum roll, please . . .)

The Solera Method

Aging the wine using a "Solera" is a laborious Jenga-like process that combines the juice from prior vintages. A Solera (Spanish for "on the ground") is a tier of barrels grouped by vintage and stacked upon one another—the oldest at the bottom and youngest at the top, sort of like a pyramid of Sherry. At specific intervals, an amount of wine from the oldest, floor-level cask is removed for bottling. This barrel is then topped off with wine from the next tier up, and so on, until reaching the top row—with no barrel ever being completely emptied. The wine created in this fractional blending process is therefore a mixture of old and young and therefore has no vintage.

Types of Sherry

Fino is pale gold in color, with a nutty aroma and almond flavor. It is ideal with tapas, seafood, cured meat, and mild cheese.

Manzanilla is dry and light with floral notes of chamomile. It is excellent with seafood, mild cheese, whitefish, and ham. It is best served chilled.

Oloroso is dry and full-bodied but can be made into a sweet dessert wine. It works before meals or with game and red meats.

Amontillado is dry, amber in color, with a caramel-like taste. It's a wonderful aperitif and pairs with white meats and oily fish.

Palo Cortado is a hybrid of Fino and Oloroso. It is a mahogany-colored wine with a hazelnut bouquet and a dry palate.

Pedro Ximénez (or PX) is molasses-dark in color and used to sweeten other sherries. It is an ultra-sweet dessert wine.

Cream Sherry is usually an Oloroso sweetened with Pedro Ximénez. Cream sherries are generally sweetened with Amontillado or Oloroso.

Pale Cream is pale, delicate, and slightly sweet. It pairs with salty foods, nuts, and olives.

Medium is amber to mahogany-colored with pastry notes and a touch of sweetness.

Sherry can be used in cooking and in cocktails, sipped before and after meals, and has true quality-to-price-ratio value. And yet, to many of us, the word "Sherry" conjures up images of your grandmother sipping a glass of her sticky-sweet favorite Sherry or an episode of the sitcom *Frasier*. Maybe it's time to rethink and re-*taste* Sherry.

85. HOW MANY KINDS OF "CHAMPAGNE" ARE THERE?

The name "Champagne" is so branded that it has become a catch-all for all sparkling wine, much to France's great aggravation. The French do have a point–Champagne is a place. And the wine must be made in that place. Look at it this way: A Jersey tomato should have to come from New Jersey, shouldn't it?

Location isn't the only legally protected aspect of Champagne. The way Champagne is made, *méthode champenoise*, is also legally protected. The process includes the addition of *liqueur de tirage* (a liquid solution of yeast, wine, and sugar) to the still wine, which causes a second fermentation to happen right inside the bottle. The resulting carbon dioxide is then trapped in the bottle, and voilà! You have Champagne! *Note:* That is a simple explanation for a complicated process.

Ironically, bubbles in wine were initially considered to be a fault since early versions of glass bottles couldn't handle the pressure and many bottles exploded in the cellar.

Contrary to popular belief, Dom Pérignon did not invent Champagne. As historians allege, he *may* have said, "come quickly, I am tasting the stars," in 1667, but English physician Christopher Merrett documented how to put the fizz into sparkling wine in 1662.

Three main grape varietals are used: Chardonnay, Pinot Noir, and Pinot Meunier. A bottle stating Blanc de Blanc means it's made using 100% Chardonnay grapes. Blanc de Noir is 100% Pinot Noir. Otherwise, it's a blend.

Most Champagne is "Non-Vintage," meaning the grapes came from multiple years. If the conditions are perfect, they will make a "Vintage" Champagne and state the year on the label. That has to have 100% of the grapes grown in that year.

But perhaps the most important aspect of Champagne is that there are sweetness levels. The amount of sugar and wine added to the *liqueur d'expedition* (called the *Dosage*) determines the sweetness level. Here they are from dry to sweet:

- *Brut Nature* (Between 0 and 3 grams of sugar)
- *Extra Brut* (less than 6 grams of sugar)
- *Brut* (less than 12 grams)
- *Extra Dry* (between 12 and 17 grams). Yes, Extra Dry is *sweeter* than Brut.
- *Sec* (between 17 and 32 grams)
- *Demi-sec* (between 32 and 50 grams)
- *Doux* (50 grams)

Every country in the world makes sparkling wine. Here is a partial list, along with the grapes and production method used.

Sparkling wine (United States): In the United States, any grape can be used. Some are made using *méthode champenoise*, and some are not. As crazy as it sounds (and is), some are even called "California Champagne."

Prosecco (Italy): Made from the Glera grape using a process that was invented in 1895 and patented in 1907 by Eugène Charmat called (not surprisingly) the Charmat method, where secondary fermentation takes place in a stainless-steel tank. Prosecco is currently America's top-selling bubbly.

Cava (Spain): Made from the grapes Xarel-Lo, Macabeo, and Parellada. A Cava is made using *méthode champenoise* and is a good (and inexpensive) alternative to Champagne.

Asti Spumante (Italy): Made from Muscat Canelli grapes using the Charmat method and named after the town of Asti. "Spuma" means foam.

Franciacorta (Italy): Made from Chardonnay, Pinot Noir, and Pinot Blanc grapes in the traditional Champagne method . . . except it's made in Italy.

Crémant (France): A Crémant is a French sparkling wine made in the Champagne method but in a region other than Champagne.

Sekt (Germany): Made from wine grapes grown in other European countries but bottled in Germany using the Charmat method. Deutscher Sekt is made using exclusively German grapes.

Tale from the Wine Floor

Customer: Can I call this a Champagne?

Me: Well, you're welcome to call it that, but that's not what it is.

C: What would you call it?

M: I'd call it a sparkling wine with pineapple flavoring added to it.

C: If I poured it for a wedding, would people think it's Champagne?

M: No.

C: What do you think they would think it was?

M: I think they would think it's fizzy pineapple juice.

C: How 'bout if I put it in a Champagne glass?

M: Yeah, that should do it.

86. WHY IS CHAMPAGNE SERVED IN FUNNY-SHAPED GLASSES?

No one knows where the funny-glass thing started or who invented it. Folklore seems to have survived longer than actual facts. One story is that the *Great Gatsby*–style glasses, called "coupe" glasses, were modeled after Marie Antoinette's, ahem, anatomy in 1787. There are other supposed origins of the glass involving, let's say, the assets of Helen of Troy, Joséphine de Beauharnais (wife of Napoleon), Diane de Poitiers (mistress of Henri II), and Madame de Pompadour (mistress of Louis XV). I'm tempted to say you can't make this stuff up, but that's exactly what happened—somebody made it up! None of them are true. But hey, why let the truth get in the way of a good story? My guess is that someone added the Champagne glass to the 65-piece dining set that Louis XVI commissioned for his queen, Marie Antoinette. (That set *did* include porcelain bowls molded from her breast.)

In a curious case of art imitating life (or is it life imitating art?), German designer Karl Lagerfeld created a modern version of the coupe glass in 2007 using his favorite model Claudia Schiffer's bosom for the Champagne brand Dom Pérignon. The resulting glasses were sold along with a bottle of 1995 Dom Pérignon Oenothèque for $3,150. Since then, other designers have jumped on the proverbial bandwagon, designing female model–inspired Champagne glasses.

But the fact is the coupe glass, with its wide surface area and shallow bowl, dissipates Champagne's bubbles, takes on the warmth of your hand, and is too easily spilled. I can't think of any reason to use it, other than to re-create the classic toast from the movie *Casablanca*: "Here's looking at you, kid."

The flute glass that followed the coupe in popularity is also not the optimum glass for sparkling wine. Although it does show and keep the bubbles longer, it robs you of the aromatics of the wine since your nose stays outside the rim when drinking. In my opinion, the standard white-wine glass is best . . . but not nearly as much fun.

ale from the Wine Floor

Customer: Do you have Russian Champagne?

Me: We have quite a few Russian sparkling wines, yes.

C: But I want Russian *Champagne*!

M: I understand. I said that because Russia doesn't make Champagne; they make sparkling wine from Russia. Champagne is a place in France.

C: Oh. I guess it's a French thing. Next, do you have Russian vodka?

M: Yup. In fact, we have Russian vodkas from around the world!

C: That's like a joke, right?

M: Right.

Chapter Nine

READING

87. HOW DO I READ A WINE LABEL?

Reading a wine label can be confusing and sometimes even annoying. This country states this, that country states that, the fonts are weird, the text is too small . . . I could go on. So rather than show a generic wine label with arrows pointing to things that aren't on every bottle, let's talk about what's required by US law to be there.

- **The producer**—The winery or person who made the wine.
- **The varietal**—The grape used to make the wine, although it does not have to be 100% of that grape. You can state the varietal in California even if only 75% of that grape is used, 85% if the appellation is stated, and 95% if the vineyard is stated.
- **The vintage**—The year in which the grapes were harvested.
- **The place**—The growing region (also called an appellation).
- **The name**—It could be the vineyard where the grapes were grown or a name they chose for the wine.
- **The contents**—How much liquid by volume is in the bottle.
- **The alcohol**—The amount of alcohol by volume (ABV) is in the wine. The law permits a 1.5% variance.
- **The sulfite declaration**—If the total sulfur dioxide is 10 ppm or more, "Contains Sulfites" must be stated. If laboratory analysis determines sulfite content is below 10 ppm, no statement is required. (See #5, "What are sulfites?")

Label definitions:

- **Vieilles Vignes**—French for old vines. To some, that would mean a more complex wine. To others, it means the wine costs more.

- **Reserve/Reserva**–It sounds like a better bottle . . . and it could be. It could be the winery's best grapes. It could be aged for a longer time in the barrel. It could be from a more specific block of land. In all but a few countries, the term has no legal definition or requirement. For most bottles, it's marketing. Or it could be.
- **Estate Bottled**–The winery grew the grapes and made the wine all on-site. It is a good thing but not necessarily a guarantee of quality.
- **Cuvée**–French for contents of a vat or tank. It can have different meanings and be used in different ways. Sparkling-wine makers use the term for their first press or "free run" juice. In Champagne you will sometimes see *tete de cuvée*, *prestige cuvée*, and *grande cuvée* to designate their best blends. It can also mean a wine blended from different vineyards, different vintages, or different barrels. It has no legal definition.
- **Unfiltered**–Filtering a wine removes dead yeast cells and clarifies the wine. Some wine lovers believe the process strips the complexities of the wines. Most wine buyers don't care if it's filtered or unfiltered.

Tale from the Wine Floor

Customer: I have a picture on my phone . . . do you have this wine?

Me: I'm sorry, but I have never seen that wine before. And since it doesn't say where it's from or even what the grape is, I have no idea what it is.

C: Do you have anything that's like it?

88. WHY DON'T WINE LABELS LIST INGREDIENTS?

You would think that wine would be regulated and enforced by the Food and Drug Administration. It is not. Since Prohibition, it falls under the guidelines of the Alcohol and Tobacco Tax and Trade Bureau or TTB, which does not require ingredient labeling. They have, however, tried five times unsuccessfully to make a rule understandable to consumers and fair to winemakers. At the heart of the problem is the mysterious nature of fermentation; what goes in at the beginning doesn't necessarily come out at the end. You could simply list the ingredients as grapes, sugar, and yeast, but even that wouldn't be correct. Yeast occurs naturally in the grape, which then consumes the grape's natural sugar, which then gets fermented out of the final product.

There are, however, a multitude of "ingredients" at the disposal of the winemaker. Egg whites and other proteins are sometimes used to clarify the wine, but they don't leave enough of a trace to be listed. Mega Purple is a point of contention for a lot of consumers—it is a concentrate that alters the color of the juice. As sinister as it may be, it is made from grape skins and wouldn't have to be listed since it is made from the product itself. Sulfites and alcohol are the only two ingredient requirements that must be listed on the bottle. I imagine there will be a time when ingredients will be listed on every wine label, but for the time being, the majority of consumers sees wine simply as the juice from grapes or just aren't interested in knowing the nutritional value of what they see as a luxury product.

89. CAN I TELL WHAT A WINE TASTES LIKE FROM THE LABEL?

Yes! You can tell a lot from the label. Grab your sleuthing hat and magnifying glass and I'll explain how to investigate. Digging into where the grapes were grown is a good starting point. Grapes grown in certain countries, climates, and regions have a similar style. For example, a country at higher altitude will have more acidity.

The grape's geography can also provide other clues. Splitting the globe in half into New World and Old World will give you valuable insight. Wines from the New World (mainly the United States, Australia, and New Zealand) tend to be fruit-forward and full-bodied wines. Old World wines (mainly Europe) have more acidity, are known to be better with food, and are less "fruity." Once you've narrowed in on a country, the more specific the listed region, the better the wine. Usually.

The vintage is your next clue. How old is the juice in the bottle? The year on the bottle tells you when the grapes were harvested. As discussed in question #65, since very few wines get better with age (and you don't know how the wine was shipped and stored), I always recommend buying the most current vintage. However, know that a wine's tannins will soften with age.

Next, look at the alcohol level. This is important, as you may recall, because of the general rule that the higher the alcohol, the drier the wine. Although it is incredibly small and hard to find (here's where the magnifying glass comes in handy), the percentage of alcohol is listed on every bottle—but it's not always correct. Alcohol has taxes applied based on that percentage, so the higher the number, the more tax is owed by the winemaker. The government allows a tolerance of plus or minus 1.5% if the alcohol content is 14% or less. And there is a tolerance of plus or minus 1.0% if the wine has more than 14% alcohol. So, a wine that states 13.5% alcohol could be labeled 12%. And a wine that states 14% could actually be 15%.

Although it is still an educated guess, here is what you were able to deduce by looking at the label: "The wine is fruit-forward and full-bodied (the country), is on the dry side (the alcohol percentage), with moderate-to-soft tannins (the vintage)." It's elementary.

🍷 ale from the Wine Floor

Customer (a nun!): Excuse me, can you help me find a wine?

Me: Certainly, Sister.

Sister: It's called Apostle Red.

M: Ah . . . Sister, the wine you're thinking of is called ApoTHIC Red, not Apostle.

S: Oooh. That's different. (Long pause).

S: Never mind.

---

90. WHY IS THERE A BLACK ROOSTER ON CHIANTI CLASSICO BOTTLES?

The black rooster or Gallo Nero was the historical symbol of the military league of Chianti back in the Middle Ages. Emperor Ghibelline in Florence and Pope Guelph in Siena fought over the land between the two cities for years. Legend has it that a contest was finally agreed upon to decide the boundaries and put an end to the bloody rivalry once and for all. Two knights—one from Siena and one from Florence—would ride on horseback and charge toward each other when the rooster crowed at sunrise. Their meeting point would be the permanent border delineation between the two cities. Siena chose a white rooster, while the Florentines chose a black one. The Florentines diabolically did not feed their bird for two days prior to the big race and, on race day, the rooster crowed long *before* sunrise. The considerable head start allowed the Florentine knight to arrive just 20km (12.43 miles) from Siena's walls before meeting up with his nemesis.

Not the most heartwarming story to latch onto as a symbol of Chianti, but the black rooster silhouette that was in use since 1383 was officially adopted by the Chianti Classico Wine Consortium in 2005. They made only one change: If the emblem has a red circle, it is basic Chianti Classico, while a gold circle indicates Riserva.

This consortium requires that Chianti Classico must be at least 80% Sangiovese with no white grapes blended in. It also must have a minimum of 12% alcohol for regular wines. Riserva wines must be aged for a minimum of 24 months in barrel and three months in the bottle, and they must have a minimum of 12.5% alcohol.

🇹ale from the Wine Floor

Customer: Can you help me find a few bottles?

Me: I'd be happy to.

C: I need a wine with a dog on the label.

M: Rascal Pinot Noir and Raymond's Frenchie have dogs on their labels.

C: Great, now one with a horse.

M: 14 Hands and Wild Horse.

C: A duck?

M: Decoy.

C: A bird?

M: Goose Bay.

C: Wow, you sure know your wines!

M: Or animals.

C: What do think of these?

M: Oh, I'm a big fan of duck wines.

C: Thanks.

M: My pleasure.

Chapter Ten

TRAVELING

91. WHAT IS THE MOST WIDELY PLANTED WINE GRAPE IN THE WORLD?

It is safe to say that Cabernet Sauvignon is the reigning king of all grapes and the most planted wine grape in the world. The grape's thick skin makes it resistant to the elements and age-worthy, and it has the ability to grow almost anywhere on the globe. It is the "gateway drug" for most wine lovers and occupies the lion's share of most collectors' cellars . . . all with good reason. The blue-black berry is deceivingly small in size and yet has a high skin-to-juice ratio, which creates high proportions of phenols and tannins. "Phenols" (often referred to as polyphenols or phenolics) are chemical compounds that affect taste and color. "Tannins," coming from the skins, seeds, and stems, give the textural element that makes wine taste (and feel!) dry. Together they soften and develop an array of flavors over time.

What is not as well known is Cabernet Sauvignon's origin. An accidental breeding occurring in 17th-century France combined the red grape Cabernet Franc and the white grape Sauvignon Blanc—a fact not known until 1996. Its half-sibling is Merlot—both have Cab Franc as a "father" but different "mothers." The offspring was immediately adopted by Left Bank Bordeaux producers who loved the characteristics it added to their blend.

Although not quite the chameleon that Chardonnay is, it is difficult to know the style of Cabernet you are buying without a little geography lesson. Three main factors determine the different styles of Cabernet Sauvignon:

1. The climate in which it is grown: Cooler climates produce an herbaceous style with vegetal notes, specifically green bell

pepper. In warmer climates, cassis (black currant) and eucalyptus become evident.

2. The type of oak used (and the time spent there): American oak imparts stronger vanilla and coconut flavors, whereas European oak (most frequently French) offers textural subtlety with spicy notes.

3. Time: The amount of time spent in the barrel (and bottle) allows the slow ingress of oxygen, making wine taste less astringent and feel smoother.

It wouldn't be fair to say unequivocally that a specific country produces a specific Cabernet Sauvignon style. Let's do it anyway.

CABERNET SAUVIGNON: The World Tour

FRANCE: Dry, medium to full-bodied, medium alcohol with high acidity and tannin. Together with Merlot, Cabernet is a key component of Bordeaux blends.

UNITED STATES: It is legal in the United States to add up to 25% of another grape into a wine that is labeled "Cabernet Sauvignon."

California: Dry, full-bodied, with a high concentration of cassis, blackberry, and eucalyptus. Napa Valley is home to some of the world's greatest expressions of the grape, with prices to match, although some value can still be had in Lodi, Paso Robles, and San Luis Obispo. Styles can range from over-oaked, high-alcohol fruit bombs to elegant and balanced age-worthy classics.

Washington State: Although Cabernet Sauvignon is the most widely planted red grape in the state, it is found mostly in Columbia Valley, where it is characterized by lower tannins and a fruitier style. Many of the vineyards lie on the same latitude as Bordeaux.

CHILE: Aged for one to two years in American oak and reasonably priced, their Cabs are full-bodied with soft tannins and loads of dark fruit. They also have a distinct note of green bell pepper.

AUSTRALIA: Dry, full-bodied, and ripe but with a strong tannic structure. Leading Cab regions Coonawarra and Margaret River produce notes of bell pepper and jalapeño(!).

SOUTH AFRICA: Dry, full-bodied, with high levels of tannin and acidity. Cabernet does particularly well in Durbanville and areas around Stellenbosch.

ISRAEL: A wide range of styles, from fresh and fruity to complex, are grown in Golan Heights, Galilee, and Jerusalem Hills.

ITALY: Although grown in many regions, the wines garnering the most acclaim (and highest prices) are Cabs from Tuscany, where they are sometimes blended with Sangiovese to make what are called "Super Tuscan" wines.

SPAIN: Spain also uses Cabernet as a blending grape, providing structure and aromas to the local Tempranillo, Garnacha, and Monastrell grapes. It is sometimes known by its synonym, Burdeos Tintos.

ARGENTINA: Although lagging behind the country's signature grape Malbec, Argentinian Cabs are gaining popularity and acclaim. Grown under the influence of a warm continental climate, they have a great depth and concentration, leading to hints of smoke and dark chocolate.

CHINA: Surprisingly, the number-one varietal in China is Cabernet Sauvignon. Although the regions are still in flux and yet to be defined, the handful of wines reaching the international market show promise in Xinjiang and Shandong provinces.

Fun Facts:
- The word "Sauvignon" is believed to derive from the French word *sauvage*, which means "wild."
- Cabernet Sauvignon was the first grape to have its entire genome sequenced. Funding for the 2016 UC Davis genome project was provided by J. Lohr Vineyards.

ale from the Wine Floor

Customer: What wine tastes like Cab?

Me: Cab does! I'm kidding—Merlot shares a profile of Cabernet but is softer.

C: But I don't like Cab, so I thought I'd try something else like it.

M: I'm confused. If you don't like Cab, why do you want a wine that tastes anything like a Cab?

C: 'Cause that's the wine I always drink.

M: Maybe you should go in a different direction and try a Pinot Noir.

C: Does it taste like Cab?

M: Not at all.

C: I'll try it, but I probably won't like it.

M: If you don't, you can always go back to Cab. You know you don't like that.

C: True.

92. WHY DO AUSTRALIANS CALL SYRAH SHIRAZ?

This question is a real head-scratcher. Could it be that Australians refer to their country as "Oz," so "Shiraz" is a way of saying "Syrah from Oz"? Maybe it's because Australians give slang terms to everything. Or maybe it's because Australians are just funny. (Some old-timers pronounce the grape using parts of both words: "Shiraa!") Now that I've successfully avoided answering this question, there is a lot I can tell you about Australian wines.

It wasn't that long ago that Australia was known for beer and dessert wines, which they call "stickies." Now, it is known either for value brands that are fruit bombs with cute, furry animals on the label or ultra-premium wines that cost a week's salary. And both sell. A *lot*! In fact, Australia is the sixth-largest wine exporter in the world behind Italy, France, Spain, the United States, and Argentina, with over 2,000 producers in 60 different wine regions. Astonishingly, nearly 90% of the wineries are owned by four mega-companies and one *huge* family-owned business.

Even though Australia is a wine superpower, it has no indigenous *Vitis vinifera* grapes, the grape species that is the chief source for most of the world's wine. The majority of vine cuttings came from America or France.

Shiraz is the country's major grape. And that is an understatement. It is grown in nearly every region in Australia and makes up half of all red wine production. Yeah, I'd say that's major. So, why do Australians call the Syrah grape Shiraz? I told you . . . Australia is funny. Australia is also funny because it is never talked about without also adding ". . . and New Zealand." Why would two countries that have no real connection to each other always be mentioned in the same sentence? It's not like they are *that* close. They are over two thousand miles apart. They are also nowhere near the same size. Australia is roughly the size of the United States, and New Zealand is the size of Colorado. They also don't make similar-style wines. Yet Australia and New Zealand are the Batman and Robin of the wine world: You never see one without the other. Maybe it's simply because they are both Down Under countries.

You are forgiven if you think only of Yellow Tail when you see the Australian (and New Zealand!) aisle, but Australia gets credit for inventing some interesting and often amazing blends:

- **Shiraz with Cabernet Sauvignon** is a classic Australian concoction that combines the predominant grapes of Bordeaux and the Rhône.
- **Shiraz, Grenache, and Mourvedre** is a Rhône-style GSM, with Shiraz playing a bigger role.
- **Sémillon with Sauvignon Blanc** is also a blend that can be traced back to the Bordeaux region in France, but the Aussies pronounce it differently—so that counts.

The more plausible explanation for the link between the two countries is that both countries combined forces in war under the banner ANZAC (Australian and New Zealand Army Corps). They even commemorate the lives lost every April 25th on Anzac Day.

93. WHAT'S SO DIFFERENT ABOUT NEW ZEALAND SAUVIGNON BLANC?

New Zealand has become synonymous with one grape: Sauvignon Blanc. As with so many regions and wines of the world, you would think there is a long history that dates back to ancient times. Not so with New Zealand's Sauvignon Blanc.

In 1969 two brothers, Bill and Ross Spence, created a profile of the grape that was different from those grown in Bordeaux or Sancerre or any other place for that matter. Their brand Matua, with their initial run of 400 bottles, was the first to produce the distinctive grapefruit-and-grassy style that now accounts for over three-quarters of the wine exported by New Zealand. Additionally, Riesling, Chardonnay, Pinot Gris (for whites), Pinot Noir, Bordeaux varietals, and Syrah (for reds) play significant roles. But Sauvignon Blanc is by far the most planted grape in the country. As of 2020, American consumers bought over $630 million of New Zealand's $2 billion in sales.

Matua may well be the first, Brancott Estate and Cloudy Bay may garner critical acclaim, but the biggest name in New Zealand wine is most certainly Kim Crawford. What Kendall-Jackson is to Chardonnay, Kim Crawford is to Sauvignon Blanc. His (yes, Kim is a he) wine became so popular that he was able to sell it literally lock, stock, and barrel for $50 million. However, selling his wine meant selling his name!

It must feel a little odd to sell your name, sign a 10-year non-compete, and then go on to make a wine with a different brand name, but that's what he did. "Loveblock" was Kim's new wine, after the 10-year wait. I would rather sell my brand's name and keep my own. Then again, no one offered $50 million for mine. Anybody?

94. AREN'T ALL GERMAN WINES SWEET?

Let's face it, nothing anyone writes is going to change the world's perception that all Rieslings are sweet. But what the hell—I'll try one more time.

OK, class, repeat after me: "Not all Rieslings are sweet!"

Glad we got that out of the way. Now let's talk about their sweetness levels. (Sorry, I couldn't help myself.)

Germany categorizes their wine based on the ripeness of the grape at the time of harvest, rather than place of origin or sweetness level of the resulting wine. (You might have to read that sentence again.)

There are four categories in ascending order of quality: Deutscher Wein, Landwein, Qualitätswein, and Prädikatswein. Prädikatswein are "predicated" on the level of ripeness.

The six "ripeness" classifications of Prädikatswein are:

Kabinett (German for, well, Cabinet) is off-dry, light, and delicate.

Spätlese (German for late harvest) is richer and riper and has more body than Kabinett.

Auslese (select harvest/to select out) is concentrated and luscious with a small amount of the beneficial fungus Botrytis, which causes the grapes to decay on the vine. You wouldn't think that's a good thing, but it is. In fact, it is known as "noble rot." (See question #23, "What's the difference between sweet wine and dry wine?")

Beerenauslese (select berry harvest) uses half botrytised grapes to make a rare and intense dessert wine.

Eiswein uses grapes that were completely frozen on the vine so that when the grapes are crushed, the water can be separated from the ultra-concentrated nectar. (See question #17, "How is ice wine made?")

Trockenbeerenauslese (select dry berry harvest) are among the greatest dessert wines in the world. Interestingly, although "trocken" means dry, this is the sweetest German wine.

There are a dozen or so reds made in Germany, and they are getting more prevalent with climate change. The most widely planted and known red is *Spätburgunder*, which you know as Pinot Noir. The other eponymous red is *Dornfelder*. The name sounds like the producer but is actually a grape created in 1956 by August Heroldrebe and named in honor of Immanuel August Ludwig Dornfelder, a senior civil servant. The grape's thick skin produces a dark-red wine, low in tannins and considered "off-dry."

What makes German Riesling so great is their ability to age while maintaining their acidity. But one of the unusual characteristics in some aged Rieslings is a distinct nose of petrol. Yes, gasoline! As crazy as it sounds, many people believe it can be amazingly enticing and pleasant. The aroma coming from the molecule *1,1,6-trimethyl-1,2-dihydro naphthalene* (mercifully shortened to TDN) is six times more likely to occur in Riesling than any other grape. European researchers quantified values of detectable TDN in 2020. It takes just 0.000004 g /L for the average person to detect something other than the grape in a glass of Riesling. Whatever the reason or detection level, wine drinkers either love it or hate it. As for me, although I used to love the smell of gasoline as my father stopped in our local Esso station for a fill-up, I'm not a fan of it in my wine. I am, by no means, the majority. There are scores of Riesling fans that prize the gassy nose of a nicely aged Riesling.

95. DOES ARGENTINA MAKE THE BEST MALBEC?

Many people hear Argentina and immediately think Malbec and only Malbec. Fair enough. It is the country's signature grape. However, Malbec's origins are in France's southwest region of Cahors, where it is also known as Cot. It is one of the allowable blending grapes in Bordeaux. But winemaker Roberto de la Mota recognized the grape's potential as a single varietal rather than just a blending grape to add color. Further elevating Malbec's international respect were forward-thinking winemakers like Nicolas Catena, Jose (Pepe) Galante, Michel Rolland, and Paul Hobbs, who concentrated on the quality of the wine rather than its sales potential. Their vision transformed a country mainly known for mediocre bulk wine made for domestic consumption into the world's fifth-largest wine producer.

But have you ever tried Argentina's Bonarda? Experts can't agree on the origin of Bonarda. Some say it's from Piedmont, Italy. Others claim it's Douce Noir from France's Savoie region. California calls it Charbono. This under-the-radar grape is deep purple with aromas of black fruit, fennel, and plum, followed by complex notes of leather and dried fig.

Next on the list of should-be-known wines from Argentina is Torrontés. This aromatic white has white peach and apricot notes with salivating citrus fruit and a hint of spice (Viognier meets Gewürztraminer!). Add great Cabernet Sauvignon, Cabernet Franc, and Petite Verdot to the list of fruit-forward yet complex wines, and you start to rethink Argentina as a Malbec-only country. There may even be a time in the near future when Argentina is most known for taking Old World grapes and making New World wines.

Oh yeah! I didn't actually answer the question: Do they make the best Malbec? Well, it's hard to say categorically that anybody makes the best anything, but my opinion is yes! The best Malbecs are made in Argentina.

96. WHAT COUNTRY MAKES THE BEST BANG-FOR-YOUR-BUCK WINE?

The answer is Spain. I can't think of any other country—with the possible exception of Portugal—that produces such high-quality wine for so little money. Sensing there may be disagreement on this answer, I offer one word to prove my point: Cava! Cava is *the* affordable alternative to Champagne. Champagne costs three times (or more) as much as Cava, yet Cava is made in Spain using the traditional *méthode champenoise*—the same way Champagne is made in France. The major differences between Champagne and Cava (besides the price and location) are the grapes and the aging requirements. Cava uses a blend of three local white grapes (Macabeo, Parellada, and Xarel-lo), but the traditional French grapes of bubbly (Chardonnay and Pinot Noir) can also be added.

So, how can Spain make a cheaper, quality bubbly than France? For starters, land and labor prices are higher in France. Plus, Spanish winemakers were the first to use a unique machine, called a gyropalette, which looks a lot like a giant Rubik's Cube that turns and vibrates 504 bottles at shot. It replaces manual, two-at-a-time "riddling"—that time and labor-intensive practice of giving the upended bottle a swift twist to progressively move the dead yeast (or "Lees") farther down the neck of the bottle (as explained in question #85). Ironically, two French vintners, Claude Cazals and Jacques Ducion, invented the gyropalette.

Although they arrived a little late to the party, Piper-Heidsieck was the first French Champagne house to place an order for the "gyro," which opened the floodgates for the revolutionary machine to be used throughout the region. Nowadays, although a few luxury brands still do the time-honored "remuage" by hand for a small part of their production, the gyropalette is now considered *de rigueur* everywhere bubblies are made.

Besides Cava, there are other Spanish wines of quality that are also easy on your wallet. One reason for this is because Spain has the largest landmass under the vine. With over 2 million vine-strewn acres and 70 wine regions, it is essentially one large vineyard.

I could go on and on about Spanish wines, but instead, I'll give you a little cheat sheet. The backbone of Spanish wine is Rioja, with its signature grape, Tempranillo. Four terms describe the age of Rioja wines:

Vino Joven (or young wine) is a very basic wine with little or no aging.

Crianza is aged for two years, with at least one year in oak.

Reserva is aged for three years, with at least one of those years in oak.

Gran Reserva is aged five years, with at least two of those years in oak.

There is no general consensus as to whether more oak aging is better. My wife, Janet, for instance, doesn't like aged wines. This is fine by me since more aging costs more money!

In addition to reigning supreme as the world's best country for quality vs. price, Spain also holds a few other titles:

- It is the third-largest wine producer but the largest exporter of wine in the world. The majority of it (again, ironically) gets shipped to France.
- It is currently the world's biggest producer of organic wine.
- Interestingly, Spain's most widely planted grape is called Airen and is primarily used for fortifying wine.

So, if you wanna pop the cork without breaking the bank, go to Spain . . . or, uh, buy a bottle from there.

97. WHAT WINE GRAPE IS CHILE MOST KNOWN FOR?

Cabernet Sauvignon, the grape that is the most widely planted and most popular in Chile, is not the country's signature grape. Let's step back in history for some context. Spanish missionaries and conquistadors were Chile's original winemakers in the mid-16th century. But following the scourge of the vine-eating bug phylloxera in France, unemployed winemakers moved to Chile and brought with them their knowledge, technique, and, more importantly, their vine cuttings. The pest decimated most of the world's vineyards but never made it as far south as Chile, making Chile home to some of the oldest ungrafted vines in the world.

Although Cabernet Sauvignon remains Chile's most important grape from a sales point of view, Carménère has a more storied past. In a curious case of mistaken wine identity, Chile believed Carménère to be identical to Merlot. It wasn't until 1994 that French botanist Jean-Michel Boursiquot discovered that what they thought was Merlot for nearly 100 years was actually Carménère. The grape is a cross between Cabernet Franc and Gros Cabernet. Stay with me now . . . Gros Cabernet's parent grape is Txakoli, which itself is related to Cabernet Franc—making Carménère both child and great-grandchild of Cabernet Franc! Even Ancestry.com would have a hard time figuring that one out!

Side note: The Carménère grape is all but extinct in its homeland of France and yet is grown almost nowhere else but its adopted land of Chile.

98. WHAT COUNTRY DRINKS THE MOST WINE?

According to the Paris-based Organization of Vine and Wine, the United States is the biggest wine consumer. Since 2011 the United States has been drinking 13% of the world's wine production—some 872 million gallons. This statistic may be attributable to the sheer size of the country as a whole, however, because when it comes to consumption of wine per person, America has a lot of catching up to do. Andorra, a small country on the Iberian Peninsula, is the largest consumer of wine per person. With a population of 69,000 people, Andorran nationals drink 3.9 million liters of wine per year, which translates to 59.6 liters per person. Second place in our wine-drinking competition is awarded to Vatican City. With a population of only 800 (all of whom are either clergy or Pontifical Swiss Guards), residents of the Vatican consume 56.2 liters of wine per person annually. Back in the United States, each passing generation has been drinking more wine, led by the states of California, Florida, New York, Texas, and Illinois(!). But at a rate of only 9.9 liters of wine per person per year, it is ranked a paltry 55th in the world for per capita consumption.

99. WHAT ARE THE TOP 10 WINE MYTHS?

While I've covered most of these topics in detail throughout the book, I thought it might be helpful to reiterate the more common misunderstandings concerning wine. Some of these "facts" have been passed down through so many generations of wine drinkers that you'd think there were no doubts about them, and you'd be wrong. Here are a few long-held myths and their realities:

All wines get better with age. Not true. Most wines get worse! Over 90% of wine is best upon release.

Smelling the cork at a restaurant tells you if the wine is bad. Not true. Cork smells like *cork*! You are presented with the cork to see if it is dried out or broken. A crumbling cork *might* tell you the wine was not stored properly. And even then, the smell of cork could blow off in a short amount of time. That does not mean that cork doesn't ruin wine; sometimes it does. That is why some winemakers choose to use a screw cap. But presenting you the screw cap in a restaurant would be weird.

A bad vintage means that no wine from that year was good. Not true. A "bad year" does not necessarily result in a bad wine. It does mean the region or vineyard had its challenges due to weather conditions. Too hot, too cold, or too anything is reflected in the grapes. But the winemaker has many options (depending on the region) to compensate for a "bad year." Ultimately, a good winemaker can make a good wine in a bad year, just less of it.

The legs on the inside of the glass are an indication of a good wine. Not true. The lines that fall slowly on the inside of the glass are an indication of alcohol content (and perhaps sweetness). The thicker and slower the streams, the higher the degree of alcohol. The room's temperature and humidity, the surface of the glass, and residual sugar also play a part in the wine's viscosity. But saying you like the wine's viscosity doesn't sound as cool as saying it has nice legs.

The better wines have a deeper indentation in the bottom of the bottle. Not true. The hole in the bottom of a wine bottle, called a punt, is solely the result of glassblowing techniques back in the day. It prevented the bottle from having a sharp corner and made it easier to stand upright. The only reason they have a punt these days is to make buyers think it's a better wine.

Opening the bottle lets the wine breathe. Not true. Well, *kinda* not true. Yes, the components of a wine change when it is exposed to air, in the same way cutting an apple begins to turn the exposed flesh brown. The unexposed part of the fruit remains unchanged. Simply opening a bottle will have virtually no effect since the surface exposed to air is about the size of a dime. You would have to decant the wine and thereby expose *all* of the juice to oxygen.

All Rieslings are sweet. *Nein!* This assumption had validity at one point, but Riesling's modern-day styles can be dry, semi-sweet, or sweet. Even the German terms used to describe the wine (Kabinett, Spatlese, Auslese, etc.) are indications of the ripeness of the grapes and not necessarily residual sugar. As in all wines, balance is the key. If sweetness is balanced with acidity, the perception of sugar won't be as much of a factor.

A screw cap means the wine is cheap. Not true. Yes, it's true that many inexpensive wines use a screw cap. However, there are expensive wines that also do. It is the winemaker's choice. Some wineries will use one or the other depending on the varietal. A wine with less age potential, say Pinot Grigio or Sauvignon Blanc, may simply require a screw cap. In contrast, other wines, let's say Bordeaux or Barolo, may benefit from a little oxygen that the naturally porous cork allows.

Organic wine does not have sulfites. Not true. Sulfites occur naturally in all wines. A wine that is Certified Organic means that no *additional* sulfites were added.

All sommeliers are wine snobs. Not true. (Caveat: Some are but don't know that they are.)

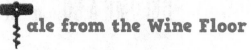

ale from the Wine Floor

Customer: Can you tell me where your good wines are?

Me: Well, a good wine is a wine you like. What kind do you like?

C: Actually, I don't like wine.

M: In that case, we only have bad wine.

100. HOW DO I BECOME A SOMMELIER?

First of all, you should ask yourself why you want to be a sommelier. If it's just for bragging rights or because you think a somm tastes wine all day, I would say you are going into it for the wrong reason. If, on the other hand, you have a passion for wine, love helping people, and you are frankly driving all of your friends crazy with constant talk about wine, then maybe you should put a cork in it and start studying.

To give you a clearer idea of the knowledge and skills required of a somm, and to see what it's like to actually take "the test" to become certified, I'm happy to share my story. It was quite the experience! Here goes . . .

I was already working in the industry and wanted to see how much I really knew about wine. So, along with my friend and co-worker Gino, I took the Court of Master Sommeliers' intro-level test. We both passed. But the way it works is, you're not technically a sommelier unless you get to the next step, the certified level. Gino was pushing me to go for it with him. "C'mon man, we got this!" he said.

I discussed the difficult decision with my wife, Janet. "It's very expensive," I said. "I'd have to do *tons* of studying. And the pass-fail ratio is terrible."

Her answer was simple. "So, you'll just have to pass."

The test is given in different cities a few times a year. Ours was set for July 30 in Boston—just five months away. Upon signing up, I received access to Guildsomm.com for one year, which has study aids and videos for the wine student. They also give you a reading and book list, but there is no course or textbook. You are pretty much on your own to study everything there is to know about wine . . . but *also* Sake, spirits, 100 classic cocktails, beer, and water! The point is that a somm should know all there is to help and guide a restaurant guest, no matter what their drink preference. As insanely broad as this is, I decided to jump in. I'd wake up each morning at 5:30, study from 6 to 9, then go to work. At night I'd run through flash cards with Janet. I searched for practice tests online. I found a lot of potential questions but no actual test. The Court is vigilant at keeping all prior tests offline. So, knowing what I needed to study more was just a guess. I never did feel confident that I was ready.

Gino and I booked a room at a hotel a quarter mile from the test location. When the time came, we drove to Boston with my car filled with cases of wine, mostly sparkling, so we could practice the service portion. We read each other's flash cards for the entire six-hour drive. After we checked into the room, we studied out loud until midnight. At 3:30 a.m. I noticed Gino was still awake, so we turned the lights back on and kept going until 5:30. We knew that rest was important, but it was impossible. At 6 a.m. I put on my new black Kenneth Cole suit and gold tie and my ultra-shined shoes and went for breakfast. We were ready. We walked the quarter mile without saying a word. The test was scheduled for 8 a.m. at a beautiful restaurant, an old and stately steakhouse with chandeliers and dark wainscoted walls. Each candidate came in quietly. Sixty-eight of us were taking the test. We were asked to form a line and show ID.

There would be three parts to the test: The first is a blind tasting of one red and one white wine; then a 40-question test with multiple choice, fill-in-the-blank, and "service." We were each told when our service portion was scheduled. To most candidates, the service portion is the scariest part. (You'll see why.) For service, you are not told until you step up whether you will be serving Champagne, a still wine, or decanting over a candle. (That's an old-fashioned method of serving old vintages, especially Bordeaux or Port. You hold a candle under the neck of the bottle to see when the sediment starts, and you stop pouring.) There could be as many as eight guests to serve. They will be hammering you with questions, asking for wine recommendations, food pairings, cocktail recipes, and origins of aperitifs. This is all happening at the same time that you are trying to do the entire "dance." And it all must be done *exactly* the correct way, in the correct order, the correct pour, the correct everything.

It was time. All of the candidates walked into a room full of round tables, silently found our seats, and watched as the four Master Somms walked in. *Note:* At the time, there were only 207 Master Somms *in the world* and fewer than 300 that *ever* passed the test to become a Master Sommelier.

"You will have 45 minutes to complete both the tasting and theory," said the proctor. "Good luck. The time starts now." I grabbed the white wine, gave it a swirl and sniff, and started filling out "the grid." The grid is a deductive tasting paper that, if properly followed, should lead you to the correct varietal, the region, and possibly even the vintage. It's amazing how blank your mind can go under stress. I could barely taste or smell *anything!* But I kept going and gave my conclusion . . . a Pinot Grigio. Italy. Trentino. One to three years. Then to the red. I thought it was a Cab from the nose, but that is the wrong way to do a blind tasting. You're supposed to stick to the grid. It should point you to the varietal and age. So, I did, but it was still a Cab. Napa. Three to five years. Again, searching for descriptors to support my conclusion, my mind went blank. But I had to move on.

I turned the paper over and began. The first five questions seemed easy enough and I felt slightly relieved. I took a deep breath and felt confident for the first time in months, even a little happy. *I got this,* I thought. But then each question got harder, and I started to sweat . . . and I don't sweat much. *Match this grape with this region. What is the French word for this and the Italian word for that.* A question on Scotch. A few questions on classifications and viniculture and soils and . . . then I heard the proctor say, "10 minutes left." I evaluated my answers. I knew I needed 28 correct to pass. I was sure of only 26. It was that close. I fought off feeling depressed as I handed in the papers and walked down the wide staircase. Everyone looked shocked. I could hear a few people quietly asking, "What did you call the white?" The way it works is, you don't *have* to call the correct wine, but you *do* have to use the correct descriptors for the wine you did call. I wasn't sure of either. But I thought I got close.

Next was the dreaded service portion. My assigned time was 10:30, with the second group of four. We were ushered into the hall. The proctor warned us with the phrase, "This is Fight Club! No talking to the others after this." I was pointed to a table and told that I would be serving a table of six. "They are celebrating with a '96 Krug. They would like other recommendations on wine for their entrees. You have 15 minutes, starting now."

I walked confidently to the table and saw that there were two men. It said "lady" at each empty seat. I greeted them with a smile (always standing to the right of the host). The somm that proctored the test was the host at my table. As the host, he was to be served the sample taste. You then have to serve the guests going clockwise. First the ladies, then the gentlemen, then back to top off the host. Always clockwise. (It's helpful for servers working the floor to know that the somm will always be going clockwise).

I walked to the guéridon (that's what that little trolley table on wheels is called) and folded three serviettes (that's what the white towels are called). They are folded ends to the middle then in half again. The first serviette goes in the ring of the ice bucket to wipe off the bottle before presenting it to the host . . . which is to be "framed" by the second serviette. A third serviette is placed on the tray, along with the perfectly polished glasses. I walked (clockwise) to place them on the table. They are to be put directly up from the fork. If a water glass is there, that should be moved to the center and the flute put in its place. I then went back for the ice bucket. That should be placed a short distance from the host: not too far that he can't pour himself, but far enough that you don't have to pour over it to serve him. I wiped the bottle, framed it to show him, and repeated the selection. "Your 1996 vintage Champagne from Krug." (It was actually an Asti Spumante!) He nodded and I got out my corkscrew. I made a nice cut and peeled off the cap (you're not supposed to use the plastic tab to take off the cap). I put it in my right pocket, along with my corkscrew, folded my serviette over the bottle, and put my thumb on top of the cork. At this point, if your thumb comes off, *you fail!* It is as simple as that. I kept it on, pulled down the metal tab, and turned six times (*every bottle* takes exactly six turns to free the cage!). You do not take off the cage from the cork . . . you hold them both and turn the bottle. The goal is to not make any sound at all. I did pretty well. Not total silence, but OK.

Then . . . *shit.* I realized that I did not place the two underliners (that's what the coasters are called) on the table, one of which is where the cork is supposed to be presented. I calmly walked to the guéridon and retrieved them. I put them down, presented the cork, wiped the top of the bottle, and poured the one and a half ounces for the host to taste. He approved, so I went to the first "lady" and around and around. Each pour has to be a steady stream without multiple stops and starts. Each glass should be filled the same. The bottle should never touch the glass. You are also supposed to calculate each pour so that a little is left in the bottle, no matter how many guests there are (so that you are not rudely forcing the host to buy another). I got close enough on it all, but I did clink one glass.

Questions were rapid-fire: "What's the origin of Campari? What's a good pairing for shrimp with saffron couscous? What's in a Godfather . . . a Madras . . . a Side Car . . . a Dark & Stormy?" Thankfully, I knew them all. I had memorized the required classic cocktails. On the one I wasn't sure about, I calmly said that I'd be happy to double-check with the bartender and get back to them as soon as I was finished serving their Champagne. The host smiled. This was an empty restaurant. There was no bartender downstairs. I was playing the game.

Then they popped one on me. "Two others will be joining us," they said. "Can you pour for them before they arrive?" I now know they were trying to throw off my calculation for each glass. Thankfully, I was able to pour the extras successfully at the guéridon. Whew! This was the end, or so I thought. They then asked me to take the table's still-full Champagne glasses away and carry them to the other side of the room and back—an odd request and not an easy task. I didn't notice anyone else doing that. But I did it. One last question: He asked if I could recommend a botrytis dessert wine. I suggested a Jurançon. He asked where that was from. I told him the southwest of France. And the grapes? Petit and Gros Manseng. He said thank you and I was done.

I walked out and felt an immediate sense of relief that it was over. All the studying, the early mornings, the flash cards. All. Over. I knew I made mistakes, but I told myself that I passed. I refused to think otherwise. I remembered our mantra, "We got this!"

Gino and I waited at a nearby Panera Bread along with the other candidates. They had come from all over the country and Canada. They looked as miserable as we did. At 3:30 we returned to the restaurant. After being ushered into the room, we were told that more than the average did not pass. *Oh great, that's encouraging!* I thought.

Glasses of Champagne were lined up on a side table with certificates and pins next to them. The first person was called: "We welcome to the Court of Master Sommeliers Mister Eugene Fitzgerald." Gino's body jerked forward, and I grabbed his hand and shook it hard. Then name after name was called in the same way . . . we welcome this person and that, but not my name. I looked at the certificates to see how many were left. There were only two. I turned away and braced myself for disappointment. And then I heard, "Welcome to the Court of Master Sommeliers Mister James Quaile." I couldn't believe it. Gino grabbed me and said words that somms shouldn't say. I walked to get my certificate, pin, and glass of Champagne. I rushed into the hallway to call my wife and kids, who were waiting (and praying). After my call, I went back in. Only 13 people passed. The ones that failed looked stunned, standing there with their wives and husbands. Some were crying.

I went over to shake hands with the host from my service portion. Knowing I made a couple of mistakes (Jurançon is not botrytis). I had to ask.

I said, "I realized I made a few mistakes."

"Yes, but it's about how you recovered," he said.

"And why the walk across the room with the glasses?"

"We couldn't rattle you," he said. "Dude, you owned it."

As we drove home that night, I relived the day's events and recalled the months and years that led me to this day. I realized for the first time that from now on, when someone asks what I do, I'd be able to say, "I'm a sommelier."

Janet waited up. We drank Scotch.

The Lightning Round

Ask a Somm

Do you drink wine every day?
There is an occasional day that I have a beer, but otherwise, yes.
What is your favorite wine (red and white)?
Red: Bordeaux. White: Chardonnay.
What was your epiphany wine?
'76 Heitz Martha's Vineyard Cabernet.
What wine do you not like?
Retsina.
What wine do you like but rarely drink?
Zins.
How do you save leftover wine?
I don't.
What is your favorite wine-and-food pairing?
Champagne with a fried oyster po'boy sandwich.
Dessert: Ruby Port with GOLDENBERG'S® PEANUT CHEWS®.
What is a grape that should be more popular?
Chenin Blanc.
What was the price of the last bottle you bought?
At a wine store: $34.99. In a restaurant: $80.
What is a wine region that overdelivers?
The Rhône.
What wine is overrated and overpriced?
Napa Cabernet.
What wine is your favorite wine to give as a gift of under $50?
Champagne.
Over $50?
Champagne.
What is the most you ever spent on a bottle of wine?
$200.
What is the one thing you wish people would pay more attention to?
Serving temperature.
Who would you most like to share a bottle of wine with?
Paul McCartney.

The Finish

I have this recurring dream, which I admit only wine nerds would understand: I am standing at the Pearly Gates, and St. Peter asks me why I left so many great bottles of wine unopened. "You were waiting for the perfect day, right? None of the days you were alive were good enough—you were holding out for a better one! Is that it?"

This dream should be a nightmare for wine collectors, who often gamble the immediate joys of today for the possible joys of tomorrow. Too many of us are saving prized bottles—telling ourselves we are waiting for *them*, when in fact, they are waiting for us!

I hope you think about my dream the next time you are with loved ones and are deciding which bottle you should open. Remember, *your* "vintage" was the rarest of all time . . . they made only one of you. And you are the best version of "you" ever made. I hope that when you reach the end of your shelf life and are asked if you opened your best wines with the ones you love, you can say, "I absolutely did, and I enjoyed every sip!"

Life is short. Open the good stuff now.

—James Quaile

Geek Speak

Wine Words & Their Definitions

ACIDITY: Tartness. Think of what a lemon does to your mouth. OK, not *that* much, but you get the idea. Ph is the measure of acidity vs. alkalinity of any liquid, on a scale of 0 to 14, with seven being neutral. The higher the pH, the lower the acidity. Typically, wines from colder climates have higher levels of acid. New wine drinkers don't care about the term.

ANGULAR: A wine with notes that hit you on different parts of your palate. An angular wine is usually high in acid. It isn't used very often. It shouldn't be. It sounds pretentious.

AROMA: The smell of the grape. The smell of the wine is the bouquet.

AUSTERE: Wines with less fruit and higher acidity. Not always a bad thing. Sometimes wines with less fruit are more in balance with all the other elements of the wine. The term is used too often by wine professionals and not at all by everyday wine drinkers.

BALANCE: When all the elements of wine come together. Think of a band singing harmony: when that one guy (it's always the tenor) is singing louder than the other guys. All of the parts need to be singing at the same volume.

BARNYARD: It is what you think it is—an aroma that reminds you of a barnyard. As insane as this sounds, it does add an intriguing complexity to wine in small doses. But too much reminds you of a barnyard!

BODY: Having to do with alcohol level, it is mainly felt on the palate and is best described like milk; Light-body = Skim, Medium-body = Whole, Full = Cream. It's the weight of the wine in your mouth.

BOUQUET: The total smell of the wine. The smell of the grapes is the *aroma*.

BUTTERY: Rich. Creamy on your tongue. Low acidity with a smooth finish. Like buttah!

CLARET: A term mostly used in Great Britain referring to a red Bordeaux wine. I've read that it was a melding of two words, clear and red wine. I dunno. Maybe.

CLOSED: A wine that is possibly good but isn't showing its full potential. It's giving the wine the benefit of the doubt, but it doesn't taste like *anything* at that moment. Not yet "open."

CLOYING: A wine that is overly rich or too sweet, without enough acidity to offset.

COMPLEX: A well-balanced wine with layers of flavor that change with time and has length on the finish. A lot going on in the glass.

CORKED: A wine that smells like musty, wet cardboard or cork! It can sometimes blow off, but it usually doesn't.

CREAMY: This term is most associated with Champagne and a malolactic fermentation process, which makes that big American Chardonnay feel so buttery on your palate.

CRISP: Stemming from acid levels, it is a term most often used to describe white or Rosé wines. A wine that makes you salivate.

CUVÉE: An unregulated word, mainly used in Champagne to denote a specific blend or reserve (which is also unregulated). Winemakers use it because it sounds classier than the literal French translation, tank.

EARTHY: Think of all elements in the earth—leaves, rocks, dirt, mushrooms, potting soil. It can be a good or bad trait, depending on the level of earthiness you like—or how grounded you are.

ELEGANT: A wine that isn't over-the-top. Not overly fruity, overly acidic, or overly anything. It is generally considered a good thing, depending on your preference. The term is most often used to describe French Burgundy. OK, it's *always* used to describe French Burgundy. I think it's a law in France.

FAT: It would be easy just to list synonyms for the word fat, but in wine it means there is a lack of acid to make the fruit taste fresh and vibrant. Like in society, it's not a desirable term no matter what the context.

FLABBY: A wine having no acidity. Just fruit. Like flat soda.

GREEN: The fruit seems underripe or literally has a vegetal aroma and taste.

HERBACEOUS: Pretty much what you think it means. Smells like herbs.

HOT: A wine with overly high alcohol levels.

JAMMY: The term is usually applied to red wines with high alcohol and low acidity. Grape Jam, get it?

JUICY: Silly term. Wine is juice.

LEES: Leftover yeast particles that sink to the bottom of the fermentation vessel as the wine is being made. The winemaker will sometimes stir the lees (called *bâtonnage* in French) to increase extraction and give a creamier texture.

LEGS: We're still talkin' wine here. After swirling, they are the long streams running down the inside of the glass—mainly an indicator of higher alcohol.

LENGTH: The amount of time a wine lingers on the palate after swallowing, a.k.a. the "finish." A longer length typically indicates greater complexity. Talking about it is OK. Timing it on your watch is pretentious.

MIDPALATE: The midpalate refers to the flavor and texture that develops in the mouth after that first taste. A wine that gets better before you even swallow is said to have a good midpalate. Wine professionals will sometimes swish and swirl the wine in their mouth to draw the flavor out further—looking like a five-year-old playing with milk.

MINERALITY: An increasingly overused term. Smells and tastes like stones, slate, wet concrete, or chalk. It sounds weird but can be really good. There is considerable debate as to whether the mineral character actually comes from the soil or not . . . or if it exists at all! If you'd like to see a customer's eyes glaze over, start talking about minerality.

MUSELET: The wire cage that fits over a a bottle of sparkling wine. Yeah . . . it has a name!

OAKED: Amazingly, the smell and taste of oak got to be "normal" for wine. Although the smell obviously comes from the barrels, the origin of the oak and the age and previous use of the barrel bring different elements to the wine.

OPULENT: Big, rich, and bold.

OXIDIZED: A wine that was exposed to too much air. It is a good or bad term depending on the wine. Although most often used to describe a flaw, some wines, like Sherry and Madeira, are intentionally oxidized to give them a nutty taste.

PLONK: A derogatory term for cheap wine. You have to be over 60 years old to use the word.

RESERVE: A few countries have rules for its use (e.g., Italy, Spain), but more often than not, it has no legal meaning.

ROUND: A wine with soft tannins and low acidity tends to have a softer, more mouth-filling texture and richness of taste.

STEELY: A high-acid and bracing wine. A term generally used for un-oaked wine.

STONE FRUIT: Aromas and tastes associated with pitted fruit, such as peaches and apricots.

STRUCTURE: A wine with high tannin and acidity.

SUPPLE: A term used to describe smooth and silky wines—or car seats.

TABLE WINE: It would seem a funny term, but it actually means a wine made with grapes grown outside of regulated regions or by unapproved methods.

TANNINS: You know when you taste a very strong tea before adding any milk or sugar? That dry-mouth feeling? *That!* It comes from the skins, seeds, and stems. Undesirable (to some) in young wines but gives the wine structure and is required if the wine is going to have any aging potential. Over time, tannins evolve to the silky liquid prized by wine collectors. It's also the thing that sweet-only wine lovers hate.

TERROIR: It means a sense of place. It's an overused French term that means any and all factors that influence the character of the wine: soil, climate, elevation, what the winemaker had for breakfast . . . everything.

TIGHT: A wine that's not ready to drink or needs time in a decanter to open up.

TOASTY: Wine barrels are charred or toasted. That char is transferred to the juice. It can be a very pleasant attribute, especially with Champagne.

TREE FRUIT: Aromas and tastes associated with fruit that grows on trees, such as apples, pears, and plums. I know what you're thinking: Doesn't all fruit grow on trees? Let it go.

TROPICAL: Aromas and tastes associated with fruit from the tropics such as pineapple, mango, kiwi, lychee, and passion fruit.

UNCTUOUS: To a wine geek, it means oily or creamy, and sometimes super sweet. It's most often used to make the listener think the taster knows wine.

UN-OAKED: A wine that wasn't fermented in oak. Duh!

VARIETAL: The grape variety. The opposite of a blend.

VEGETAL: The aroma or taste of veggies. Most often green bell pepper.

VENDANGE: A French term for grape harvest. By extension, a new term officially recognized in 1984 and mainly used in the Alsace region is Vendange Tardive, meaning "late harvest."

VINTAGE: The year the grapes were grown.

YIELD: The amount of grapes harvested in a particular year.

Tale from the Wine Floor

Customer: Can you help me? You did last time I was in.

Me: I'd be happy to. With so many wines here, it feels good to help people. And yet, I don't often know how they enjoyed my recommendations. So, I just want you to know that you made my day by simply saying that I've helped you, and I'm so glad you came in again. So . . . what can I help you with today?

C: Where are your restrooms?

In Gratitude

You can't go it alone. It's true in life, and especially true when you're writing a book. Here are some of the people who were with me on this long journey.

I'd like to send a very special thank-you to my very own personal editor, Maureen Busarello, who said things only a sister could say . . . because she *is* my sister! Her thoughts and edits made the book 100 times better, and her humor made it fun to write. Thanks, Reen!

I loved wine but didn't really understand it until I began having tastings with Jim Auletto and Joe Brandolo. Both are generous men who will share their knowledge (and wine!) with anyone who shares their passion.

Joe Arking has been a game-changer for so many people, myself included. My life in the wine industry is due to him and his generosity.

Jamie Arking is my day-to-day superior at work, but you'd never know it. He treats me—along with everyone else—with respect. He's a mensch!

I'd like to thank a few of my many wine friends for their valued opinions and support: Joe Lynch, Keith Bader, Phil Tartaglione, Gino Fitzgerald, Matt Hedges, Bryan Robey, Frank Catrombone, Jean-Louis Carbonnier, Mark Steinberg, and Joe and Pat Nero. Cheers!

I am much obliged to my fact-checking fellow wine geek Ted Lippman for his help and encouragement.

Thanks also to those who helped put me on the path to publishing: Lindsay Newton, Dominick Nero, and David Litt.

I would like to offer a long overdue debt of gratitude to my high school English Lit teacher and lifelong friend Kenneth Kaleta. I hope he sees his wit in my book. It was my intention.

I am hugely thankful for my agent, Jackie Meyer. She saw what others didn't and made it happen when others couldn't.

I am so grateful for Rick Reinhart. His team at Globe Pequot and Rowman & Littlefield, especially Felicity Tucker and Marianne Steiger, pulled out all the stops on the way to bringing this book to publication and Alyssa Griffin handled the marketing. Obviously, I couldn't do it without them!

To my friend of 40 years John O'Brien, I would like to express my gratitude for his support from the very beginning. I am so proud to feature his art in my book.

Lastly, I must toast my wife, Janet. I can't name a time when I proposed a new venture that she didn't encourage me to go for it. I am happy to be traveling this long and winding road with her by my side.

Index

Beaujolais: Beaujolais Nouveau compared with, 139; in bistro, 2; classification system of, 136; Gamay grape and, 139; similar wines to, 47; at Thanksgiving, 91–92, 139

Beaujolais Nouveau, 139, *140*, 141

Bible, wine mentions in, 132

biodynamic wine, 20

black rooster, on Chianti bottles, 168

blending: benefits of, 78; of Rosé wines, 50

blends, 78; in Australia, 175; in Bordeaux, 135–36

blind tasting: Judgement of Paris, 47; in sommelier exam, 5, 189–90; theory of, 46–47

body, of wine, 199

Bonarda wine, 179

Bordeaux, France, 30, 66, 135–36, 143, 195

Bordeaux wines, aging and, 135

Botrytis (Noble Rot), 36, 192

Bottle Shock (film), 47

bouquet, of wine, 199; aroma contrasted with, 38

Boursiquot, Jean-Michel, 182

box wines, 76

brandy, 154

bring-your-own-bottle party (BYOB), 102, 129

Brix level, 32

Burgundy, France, 66, 136

buttery wine, 199

buying, of wine, 56–85, 129–30

BYOB. *See* bring-your-own-bottle party

Cabernet Sauvignon wine, 68, 108, 170–72, 195

cage, of sparkling wine bottle, 191

Cahors region, of France, 179

California Sustainable Winegrowing Alliance, 21

Canada, 26

carbon dioxide, 8

carbonic maceration, 139

Carménère wine, 182

Casablanca (film), 161

case of wine, size of, 123

Catholic Mass, wine production for, 84

Cava, 71, 159, 180

Certified Sommelier exam, 2–3, 5

Champagne: flute, 161; other sparkling wines contrasted with, 180

Champagne, France, 33, 132, 135, 158

Champagne glasses, 161

Charbono wine, 179

Chardonnay wine, 136, 195

Charles Shaw wine, 65

Château d'Yquem wine, 117

Chenin Blanc wine, 195

Chianti, Italy, 132

Chicago Saint Valentine's Day Massacre, 84

Chile, Cabernet Sauvignon in, 171, 182

China, 172

cigarette smoking, tastes and, 52

Claret, 143, 199

clarification agents, 18, 59

Classico DOCG, 145

classifications, for Italian wine, 144–45

Clef du Vin device, 116

climate change, wine and, 30–31

closed wine, 200

cloying wine, 200

cocktails, in sommelier exam, 191

collections, of wine, 129

colors, of wine bottles, 115

commodity, wine as, 57

common mispronunciations, 150

complex wine, 200

consequences, of Prohibition, 85

consumption of wine, time period after buying of, 116

contact with juice, of grape skins, 7

convenience, of box wine, 76

cooked wine, 112

cooking wine, 74–75

cooling, of wine, 96

Coppola, Francis Ford, 149

Coravin device, 108

cork: breaking of, 110; as porous, 121; presentation of, 63, 185; pushing in of, 110; reasons for use of, 118; as sustainable, 119

cork alternatives, search for, 119

corked wine, 113, 118, 119, 200

Côtes du Rhône wine, 92, 137, 150, 195

coupe glasses, 161

Court of Master Sommeliers, 5, 188–89

cow horn procedure, in biodynamic farming, 20

Crawford, Kim, 176

Cream Sherry, 157

creamy wine, 200

Crémant sparkling wine, 159

Crianza Rioja, 181

crisp wine, 200

crown caps, Pét-Nats and, 23

Cru Beaujolais, 141

Cru growing sites, 141–42

culture, familiar flavors and, 41

customer preference, substitutions for, 47–48

cuvée, 200

Daily Value (DV), 39

decanting, 11, 95; over candle, 189

deductive tasting, 46

deglazing, 74

Denomination of Origin System (DO), 155

Denominazione di Origine Controllata (DOC), 145

Denominazione di Origine Controllata e Garantita (DOCG), 145

dessert wines: Australian, 174; legs of, 43

different tastes, of same wine, 51

Digby, Kenelm, 122

dinner guests, method for serving wine to, 190–91

dishwasher, wineglasses and, 101

distilled spirits, addition to wine of, 152, 154

distributors, of wine, 84–85

diurnal shift, 30

DO. *See* Denomination of Origin System

Germany, 26, 31, 177
Gewürztraminer wine, 150
Giamatti, Paul, 53
Gibbs-Marangoni effect, 43
gifts, wine as, 68–69
Gironde River, 135
glassblowers, wine bottle size and, 122
glass of wine bottle, thickness of, 124
glassware, recommendations for, 100
Glühwein drink, 93
gluten-free wine, 60
The Godfather (film), 149
Goldilocks zone, of wine, 96
good value wine, 180–81
grams per liter measurement, of sugar, 32
Gran Reserva Rioja, 181
grapes: allowed in Bordeaux, 135–36; amount of required, 10
grape skins: contact with juice, of, 7; development of, 30
Graves Region, of France, 117
Great Depression, 84
green wine, 200
guests, for wine tasting, 103
gyropalette machine, 180

Hall, Allan, 139–40
harvest date, 8, 30, 66
Harvey's Bristol Cream, 1
headaches, red wine and, 13–14
health, wine and, 39
heat-tolerant varietals, 31
herbaceous wines, 54, 200
Heroldrebe, August, 178

high alcohol, sugar consumption and, 32
higher ground, for grapes, 31
histamines, 13–14
holidays, dedicated to wine, 106
horizontal tastings, vertical and, 44
hot wine, 200
hybrid varieties, heat tolerance and, 31
hydrogen peroxide, sulfites and, 14

ice buckets, 97
ice wine, 26, 36, 178
IGT. *See* Indicazione Geografica Tipica
INAO. *See* Institut National des Appellations d'Origine
indentation, at bottom of wine bottle, 27
Indicazione Geografica Tipica (IGT), 145, 148
ingredients, on wine labels, 1, 165
in perpetuum process, 153
Institut National des Appellations d'Origine (INAO), 133
investment collectors, of wine, 129
Italian wine, 144, *146*; classifications for, 145
Italy, 66, 144–45, 172

jammy wine, 200
Jewish Dietary Laws (Kashrut), 25
Judgement of Paris tasting, 47
jug wine, 84, 109
juicy wine, 200

National Prohibition Act
(Volstead Act), 83–84
natural wine, 22
nature, influence on taste buds
on, 41
negociant (wine merchant), 139
New Jersey, 1
new smells, brain processing of,
41
new wine regions, 31
New World wine, Old World
contrasted with, 48, *48,* 166
New Zealand, 174–75
Noble Rot (Botrytis), 36
now or later approach, to wine
buying, 129–30
nurture, influence on taste buds
of, 41
nutrient contents, of wine, 39, *40*

oak barrels: reasons for, 127;
surname "Cooper" and, 128
oaked wine, 1, 57, 127, 128, 201
oak trees, 127
oenophiles, 35
oil-soaked rags, wine closures
and, 118
Old World wines, New World
contrasted with, 48, *48,* 166
Oloroso Sherry, 157
opening of wine bottle, breathing
of wine and, 186
Oporto, Portugal, 154
opulent wine, 201
orange wine, 24
order, for wine tasting, 104
ordering, of wine, 62
organic wine, sulfites and, 14,
19
organization, of wine list, 62

Organization of Vine and Wine,
183
Orthodox Union, 25
oxidized wine, 201–2
oxygen, wine and, 95

pairing, of wine, 87–88, 195
palate, expansion of, 41; prac-
tice for, 42
Pale Cream Sherry, 157
Paris trip, 2
Paris Wine Tasting, of 1976, 47
Parker, Robert, 64–65, 148
party: amount needed for,
70–71, *72*; bring-your-own-
bottle (BYOB), 102, 129;
types of wine for, 70
Pasteur, Louis, 7
PDOs. *See* Protected
Designation of Origin
Pedro Ximénez Sherry, 156, 157
Peter (Saint), 197
Pétillant Naturel wines (Pét-
Nats), 23
phylloxera insect, 182
Pinot Grigio, 70
Pinot Noir wine, 46, 136, 139
plastic wrap trick, 113
plonk, 202
point scoring system, for wine,
64–65
polyphenols, 11
Port wine, 75, 92, 154
preference, aging and, 181
Preparation 500, 20
preservation: kits, 108; of wine
for sea voyages, 152, 154
price, recommendations for,
62–63, 80, 195
primary flavor notes, 38

Printed in the USA
CPSIA information can be obtained
at www.ICGtesting.com
LVHW091410261023
761444LV00002B/3